ARTHUR GODMAN

LONGMAN CHEMISTRY HANDBOOK

the fundamentals of chemistry
explained and illustrated

D1465134

LONGMAN   YORK PRESS

YORK PRESS
Immeuble Esseily, Place Riad Solh, Beirut.

LONGMAN GROUP UK LIMITED
Longman House, Burnt Mill, Harlow,
Essex CM20 2JE, England
and Associated Companies throughout the world.

First published 1992
Third impression 1993

ISBN 0 582 08810 0

Diagrams by Rosemay Vane-Wright
Colour origination by Chris Willcock Reproductions
Photocomposed in Britain by Composing Operations

Produced by Longman Singapore Publishers Pte Ltd
Printed in Singapore

Contents

How to use the handbook

This handbook contains some 1500 words used in chemistry. These are arranged in groups under the main headings listed on pp. 3–4. The entries are grouped according to the meaning of the words to help the reader to obtain a broad understanding of the subject.

At the top of each page the subject is shown in bold type and the part of the subject in lighter type. For example, on pp. 12 and 13:

12 · PROPERTIES OF SUBSTANCE/CHANGE OF STATE

PROPERTIES OF SUBSTANCE/PHYSICAL PROPERTIES · **13**

In the definitions the words used have been limited so far as possible to about 2000 words in common use. These words are those listed in the 'defining vocabulary' in the *New Method English Dictionary* (fifth edition) by M. West and J. G. Endicott (Longman 1976). Words closely related to these words are also used: for example, *characteristic*, defined under *character* in West's *Dictionary*. For some definitions other words have been needed. Some of these are everyday words that will be familiar to most readers; others are scientific words that are not central to chemistry. These are contained on pp. 211–33 and will be found listed in the alphabetical index.

1. To find the meaning of a word

Look for the word in the alphabetical index at the end of the book, then turn to the page number listed.

The description of the word may contain some words with arrows in brackets (parentheses) after them. This shows that the words with arrows are defined near by.

(↑) means that the related word appears above or on the facing page;

(↓) means that the related word appears below or on the facing page.

A word with a page number in brackets (parentheses) after it is defined elsewhere in the handbook on the page indicated. Looking up the words referred to in either of these two ways may help in understanding the meaning of the word that is being defined.

The explanation of each word usually depends on knowing the meaning of a word or words above it. For example, on pp. 91–2 the meaning of *crystalloid*, *crystallization* and *polymorphism*, and the words that follow depends on the meaning of the word *crystal*, which appears above them. Once the earlier words are understood those that follow become easier to understand.

2. To find related words

Look in the index for the word you are starting from and turn to the page number shown. Because this handbook is arranged by ideas, related words will be found in a set on that page or one near by. The illustrations will also help here.

For example, words relating to crystal systems are on pp. 96–7. On p. 96 *crystal systems* is followed by words used to describe various kinds of systems and types of structures; pp. 98–100 give words for the related subject of colloids; p. 101 lists words for the properties of different types of colloidal dispersions.

3. As an aid to studying or revising

There are two methods of using the handbook in studying or revising a topic. You may wish to see if you know the words used in that topic or you may wish to revise your knowledge of a topic.

(a) To find the words used in connection with crystals look up *crystal* in the alphabetical index. Turning to the page indicated, p. 91, you would find *crystal*, *crystalloid*, *crystallization*, *recrystallization*, and so on. Turning over to p. 93 you would find *pattern*, *symmetry*, etc.

(b) Suppose that you wished to revise your knowledge of a topic; e.g. *isotopes*. If, say, the only term you could remember was *relative molecular mass* you could look it up in the alphabetical index. The page reference is to p. 114. There you would find the words *isotope*, *isotopic ratio* and *relative isotopic mass*, etc. If you next looked at p. 113, you would find words relating to the structure of the atom which would help you to understand the descriptions given for the words connected with isotopes.

4. To find a word to fit a required meaning

It is almost impossible to find a word to fit a meaning in most dictionaries, but it is easy with this book. For example, if you had forgotten the word for the substance from which metals are obtained, but you knew such substances came from a mine, all you would have to do would be to look up the word *mine* and turn to the page indicated, p. 154. There you would find the word you wanted, which is *ore*, and also related words such as *deposit*, *seam* and *lode*.

THE
HANDBOOK

material (*n*) a material has some general
properties (↓) by which it can be recognized.
Other properties can vary between different
kinds of the same material. Examples of
materials are wood, leather, rubber and brass.
Different kinds of wood have slightly different
properties: their colours vary, their densities
(p.12) vary, their hardness varies. The variation
of the properties, from one kind of the same
material to another, is small. The chemical
composition (p.82) of a material can also vary,
but the variation is small. **material** (*adj*).

different materials

substance (*n*) a substance has properties (↓) by
which it can be recognized. These properties
do not vary from one piece of the substance to
another. The chemical composition (p.82) of a
substance does not vary. Examples of
substances are: iron, cane-sugar, salt. Many
substances are compounds (↓); some
substances are elements (↓).

compound (*n*) a substance (↑) that can be
decomposed by chemical action (p.19) into
simpler substances. The chemical constitution
(p.82) of the substance is known, and a
chemical formula (p.78) can be given for it. For
example, lime is a compound of calcium and
oxygen; one atom (p.110) of calcium combines
with one atom of oxygen to form one molecule
(p.77) of lime (calcium oxide); the chemical
formula is CaO. To contrast *material* (↑),
substance (↑), and *compound*: a *material* (e.g.
wood) has a chemical composition (p.82) and
properties (↓) which may vary between limits;
a *substance* (e.g. a protein) has a definite
chemical composition, but its constitution may
be too complex to be described; a *compound*
(e.g. sulphuric acid) has a definite chemical
composition, a known chemical constitution,
and can be given an exact chemical formula.

different compounds

element[1] (*n*) a substance (↑) that cannot be
decomposed by normal chemical action (p.19)
into simpler substances. All substances and
materials (↑) are composed of elements which
are chemically combined. *See page 116 for a
modern definition of element.* **elementary** (*adj*).

different elements

gas

widely spaced particles
move very fast

liquid

less widely spaced
particles move
around

close-packed
particles vibrate

solid

three states of matter

property (*n*) a property can be seen, heard,
smelt, or felt by the senses and it allows one
material (↑) or substance (↑) to be recognized
as different from another material or substance.
All materials and substances have physical
properties (↓) and chemical properties (p.19).

physical property a property (↑) which does not
depend on the effect of other materials or
substances. Examples of physical properties
are: shape, colour, odour (p.15), solubility
(p.87), melting point (p.12), density (p.12). *See
chemical property (p.19).*

intensive property a property which does not
depend upon the amount of material or
substance; such properties are used to identify
(p.225) a material or substance, e.g. colour,
odour, density, boiling point (p.12),
concentration (p.81).

extensive property a property which depends
upon the amount of material or substance;
such properties are used to identify different
specimens (p.43) of the same material or
substance, e.g. mass, volume.

characteristic (*adj*) describes a property which
readily distinguishes (p.224) an object,
material, substance, pattern (p.93) from all
other similar things. A characteristic property
provides an easy means of recognition, e.g.
copper has a characteristic reddish-brown
colour, which readily distinguishes it from other
metals. **characteristic** (*n*).

feature (*n*) a distinctive property common to a
group of materials or substance.

description (*n*) a list of the properties of an object,
material, substance, pattern, form of energy,
or collection, or a list of the events in a process.
describe (*v*), **descriptive** (*adj*).

state of matter solid, liquid, and gas are the three
states of matter. Any material or substance is a
solid, a liquid or a gas.

change of state a physical change (p.13) of a
material or substance from one state of matter
(↑) to another e.g. from solid to liquid; from
liquid to gas. A change of state is commonly
caused by heating or by cooling.

solid (*n*) one of the states of matter (p.9). A solid has a definite (p.226) mass, a definite volume and a definite shape. For example, iron is a solid at room temperature. **solid** (*adj*), **solidify** (*v*), **solidification** (*n*).

melt (*v*) to change a solid (↑) to a liquid (↓) by heating; a solid to change to a liquid when heated. For example, heat melts ice; ice melts when heated. Only one material or substance is concerned in the action. Compare *dissolve* (p.30) which concerns two or more substances. **molten** (*adj*).

molten (*adj*) describes a material or substance in the liquid state. The material or substance is solid at room temperature.

solidify (*v*) to change a liquid (↓) to a solid (↑) on cooling. Solidifying is the opposite action to melting, only one material or substance is concerned in the action. The word is used for materials and substances which are normally solid at room temperature, e.g. molten iron solidifies at about 1500°C.

set (*v.i.*) of suspensions (p.86) in liquids, to form a solid as the liquid evaporates.

freeze (*v*) to change a liquid (↓) to a solid (↑) on cooling below room temperature. The word is used for substances which are normally liquid at room temperature, e.g. water freezes to form ice. Freezing is the opposite action to melting. **freezing** (*adj*), **frozen** (*adj*).

liquid (*n*) one of the states of matter. A liquid has a definite (p.226) mass, a definite volume, but no definite shape, e.g. water and kerosene are liquids at room temperature. A liquid takes the shape of its containing vessel (p.25). **liquefy** (*v*), **liquefaction** (*n*), **liquid** (*adj*).

boil (*v*) to change a liquid (↑) into a gas (↓) by heating. Bubbles (p.40) of gas are formed in the liquid and the gas is given off (p.41). The temperature of the liquid remains constant (p.106) during boiling. **boiling** (*adj*), **boiled** (*adj*), **boil** (*n*).

boiled (*adj*) describes water that has been boiled for some time. The water no longer contains dissolved (p.30) air.

vaporization

vapour

liquid

temperature below
boiling point

vaporization

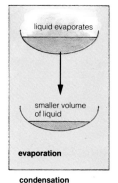

liquid evaporates

smaller volume
of liquid

evaporation

condensation

cold

vapour
condenses
to liquid

vapour

hot

liquid

liquefy (*v*) to change a solid (↑) or a gas (↓) to a
liquid (↑). A gas is liquefied by cooling.
liquefaction (*n*).

gas (*n*) one of the states of matter (p.9). A gas
has a definite (p.226) mass but no definite
volume and no definite shape, e.g. air is a gas
at room temperature and pressure. A gas
expands to fill the volume of its containing
vessel (p.25). **gaseous** (*adj*).

gaseous (*adj*) describes a substance in the state
of a gas (↑), or a chemical reaction (p.62)
between gases.

vapour (*n*) a substance in the gaseous (↑) state.
A vapour can be changed to a liquid (↑) by
increasing the pressure (p.102). A gas is called
a vapour below its *critical temperature* (p.102).
To compare *gas* with *vapour*: both are in the
gaseous state, but a gaseous substance above
its critical temperature is a *gas* and cannot be
liquefied (↑) however great the pressure, while
a gaseous substance below its critical
temperature is a *vapour* and can be liquefied
by a sufficient increase in pressure. **vaporize**
(*v*), **vaporization** (*n*).

vaporize (*v*) to change a liquid (↑) to a vapour
(↑) at a temperature lower than that at which it
boils. For example, naphthalene vaporizes at
room temperature. **vaporization** (*n*).

evaporate[1] (*v*) to change a liquid (↑) to a vapour
(↑) and so to cause the volume of the liquid
slowly to become less. The important fact is the
volume of liquid becoming less. *See evaporate*[2]
(p.32). **evaporation** (*n*), **evaporated** (*adj*).

condense (*v*) to change a vapour (↑) to a liquid
(↑) by cooling, or by increasing the pressure,
or by both; change of a vapour to a liquid
because it cools or because the pressure is
increased. The word is used for materials and
substances which are liquid at room temperature and the usual method of condensation is by
cooling. **condensed** (*adj*), **condensation** (*n*).

condensation[1] (*n*) the formation of a liquid from its
vapour, e.g. the condensation of steam to water.

fluid (*n*) any liquid or gas is a fluid. A fluid is a
substance that flows. **fluid** (*adj*), **fluidity** (*n*).

boiling point the temperature at which a liquid (p.10) boils (p.10). At the boiling point the *vapour pressure* (p.103) of the liquid is equal to the *atmospheric pressure* (p.102). The lower the atmospheric pressure, the lower is the boiling point of the liquid. The boiling point of water is 100°C at *standard atmospheric pressure* (p.102).

melting point the temperature at which a solid substance melts; at the melting point both solid and liquid substance exist (p.213) together. The melting point of a solid varies slightly with the ambient (p.103) pressure (p.102). The term melting point is used for substances which are solid at room temperature. *See freezing point (↓).*

freezing point the temperature at which a solid substance melts. The term freezing point is used for substances which are liquid at room temperature, e.g. the freezing point of water is 0°C but the melting point of naphthalene is 80°C.

mass (*n*) the property of a material or substance that causes it to be attracted by the earth. The *force* attracting an object, or any substance, to the earth is its *weight*. Mass is measured in *kilograms*, weight is measured in *newtons*.

weight (*n*) see mass (↑). **weigh** (*v*).

volume (*n*) the amount of space taken up by an object in three dimensions.

density (*n*) the mass (↑) of 1 m³ of a material or substance (p.8). Density = mass ÷ volume for any specimen (p.43) of a material or substance. Density is an intensive property (p.9) used in identification (p.225) of material and substances. Density is measured in kg/m³. **dense** (*adj*).

relative density the density of a material or substance divided by the density of water. Relative density has no units, it is a pure number.

relative vapour density the density of a gas or vapour divided by the density of hydrogen measured at the same temperature and pressure (p.102). Relative vapour density has no units, it is a pure number, and is independent of temperature and pressure. The relative vapour density of a substance is numerically equal to half its *molar mass* (p.114).

vapour density = relative vapour density (↑).

boiling point

boiling point

thermometer

vapour

boiling liquid

heat

melting point

thermometer

melting point

melting solid

heat

different densities

gold
0.1 kg
5.2 cm³

iron
0.1 kg
12.7 cm³

aluminium
0.1 kg
37.0 cm³

physical change a change in which no new materials or substances are formed. In a physical change, a material or substance may change its state, or some of its physical properties (p.9) may change, e.g. the change from water to steam is a physical change.

state of division a measure of the size of small particles (↓) into which a large piece of a solid has been divided, e.g. marble can be in lumps (↓), chips (↓), or powder (↓), three different states of division.

particle[1] (*n*) a very small piece of solid material or substance (p.8).

lump (*n*) a large piece of a solid material or substance with an irregular (p.93) shape. **lumpy** (*adj*).

chip (*n*) a small piece of a solid material or substance, broken off from a large piece. A chip is smaller than a lump, but bigger than a granule.

flake (*n*) a small, flat lump (↑) of solid material or substance. A flake is similar in size to a chip (↑).

granule (*n*) a small piece of a solid material or substance made up of several grains (↓). **granular** (*adj*).

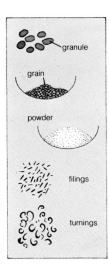

grain (*n*) a very small piece of solid material or substance; a particle (↑) that can be seen by the naked eye. Sand and salt consist of grains.

powder (*n*) a solid material or substance consisting of particles (↑) so small that they cannot be seen by the naked eye. **powdered** (*adj*), **powdery** (*adj*).

filings (*n.pl.*) small particles formed by rubbing a metal; they are similar in size to grains or granules (↑), but long and thin.

turnings (*n.pl.*) particles formed by cutting a metal; they are much larger than filings (↑).

fine (*adj*) describes powders (↑) and filings (↑) in which the state of division (↑) is very small. **fineness** (*n*).

coarse (*adj*) describes powders and filings which are bigger than those described as fine (↑). **coarseness** (*n*).

finely divided (*adj*) describes a solid in powder form with very small particles in the powder, i.e. a fine powder.

The labels on the figure read: granule, grain, powder, filings, turnings.

texture (*n*) the nature (p.19) of the surface of a solid, i.e. whether it is rough or smooth, is the texture of the surface. The texture of a powder, granules or grains, depends upon the fineness (p.13) or coarseness (p.13) of the particles, e.g. a surface can have a smooth texture or a powder a coarse texture.

massive (*adj*) describes a solid, particularly a metal, which consists of large pieces, including lumps, e.g. massive zinc consists of large pieces of zinc. Massive is the exact (p.79) opposite of finely divided (p.13).

elastic (*adj*) describes a solid which can have its shape changed by a force and which returns to its original (p.220) shape when the force is removed, e.g. a piece of rubber is elastic. The property of such a solid is its **elasticity** (*n*).

plastic[1] (*adj*) describes a solid material or substance which can have its shape changed by a force but which does not return to its original (p.220) shape when the force is removed, e.g. clay is plastic. The property of such a solid is its **plasticity** (*n*).

brittle (*adj*) describes a solid material or substance which breaks into small pieces under a force, e.g. glass is brittle, it breaks into small pieces when hit. The property of such a solid is its **brittleness** (*n*).

ductile (*adj*) describes a solid material or substance which can be drawn out to form a thin wire. Metals and alloys (p.55) are ductile. The property of such a solid is its **ductility** (*n*).

malleable (*adj*) describes a solid material or substance which can have its shape changed to a thin sheet by beating with a hammer, e.g. iron is malleable. Metals and alloys (p.55) are malleable. The property of such a solid is its **malleability** (*n*).

abrasive (*adj*) describes a material which wears away the surface of another material. **abrasion** (*n*).

refractory (*adj*) describes a solid material or substance which can be heated to a high temperature without changing its properties, e.g. some kinds of bricks are refractory. **refractoriness** (*n*).

original length

stretches

returns to orginal length

elastic solid

original length

stretches

does not return to original length

plastic solid

ductile solid

malleable solid

crystals

porous (*adj*) describes a solid material full of very small holes, through which fluids can pass, e.g. a brick is porous. **porosity** (*n*).

crystalline (*adj*) describes a solid material or substance (p.8) which has molecules (p.77), atoms (p.110) or ions (p.123) arranged in a regular structure (p.82). A crystalline substance forms crystals (p.91); a metal is crystalline but it does not form regular crystals. *See metal crystals (p.95)*.

amorphous (*adj*) describes a solid material or substance which has no crystalline (↑) structure. Glass, rubber and many plastics (p.210) are amorphous.

coloured (*adj*) describes a material or substance (p.8) which has colour, but without naming the colour, e.g. a coloured solution can be brown, blue, green, black, etc. A material or substance can be described as white or coloured, e.g. milk is a white liquid and lead (II) sulphide is formed as a black precipitate (p.30) which is considered to be a coloured precipitate.

colourless (*adj*) describes a material or substance (p.8) which has no colour, e.g. water is colourless; air is colourless. Colourless is the opposite of coloured. To contrast *white* and *colourless*: the paper of this book is *white*; glass in a window is *colourless*.

odour (*n*) that property of a material or substance (p.8) which is recognized by smell, e.g. onions have a well-known odour.

odourless (*adj*) describes a material or substance which has no odour (↑).

quality (*n*) a property of a material or substance (p.8) which cannot be measured, but is an observation made only by the senses, e.g colour, odour and texture are some of the qualities of materials and substances.

impart (*v*) to give a quality (↑) or a quantity (p.81) to an object or to share a quality or a quantity with an object, e.g. sugar imparts sweetness to a drink; potassium salts impart a lilac colour to a flame; an explosion imparts energy to a rocket.

surface (*n*) the outside part of a solid object; it has length and breadth, but no thickness or depth. A liquid has a surface where it meets the air. Examples: a brick has six surfaces; the surface of water in a cup.

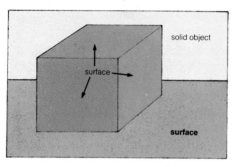

granular (*adj*) (1) describes a surface which appears to have many grains or granules. (2) describes a coarse powder consisting (p.55) of granules.

dull (*adj*) describes a surface which does not shine when light falls on it. Dull is the opposite of bright. For example, the surface of wax is dull. **dullness** (*n*).

lustre (*n*) that property of a surface which causes it to shine when light falls on it. Lustre is a quality. For example, the surface of silver has a lustre. **lustrous** (*adj*).

transparent (*adj*) describes a solid object, material or substance which allows light to pass through it, so that a person can see through it, e.g. glass is transparent. **transparency** (*n*).

translucent (*adj*) describes a solid object, material or substance which allows light to pass through it, but a person cannot see clearly through it. For example, waxed paper is translucent, but not transparent; milk is a translucent liquid.

opaque (*adj*) describes an object, material or substance which does not allow light to pass through it, e.g. leather and paper are opaque; mercury is an opaque liquid. **opacity** (*n*).

clear (*adj*) describes a liquid which is transparent, e.g. water is a clear liquid. A clear liquid can be coloured or colourless, e.g. tea is a clear, brown liquid; kerosene is a clear, colourless liquid. **clarity** (*n*).

soluble (*adj*) describes a solid, or gaseous (p.11), substance (p.8) which can be dissolved (p.30) in a liquid; the liquid is usually water. A substance can be described as very soluble, slightly (↓) soluble, sparingly (↓) soluble, insoluble (↓) or soluble. For example, sugar is soluble in water (sugar can be dissolved in water); lime is slightly soluble in water. **solubility** (*n*).

insoluble (*adj*) describes a solid, or gaseous (p.11), substance (p.8) that does not dissolve in a liquid. It is the opposite of soluble. Very few substances are completely insoluble.

slightly (*adj*) describes **soluble** (↑) if only a small amount of substance dissolves in a liquid, e.g. lime is slightly soluble in water, that is, only a small amount of lime will dissolve in water.

sparingly (*adj*) describes **soluble** (↑) if an amount which dissolves is very small, even less than dissolves for a slightly soluble (↑) substance, e.g. air is sparingly soluble in water.

flocculent precipitate

creamy precipitate

heavy precipitate

flocculent (*adj*) describes a precipitate (p.30) which has the appearance of wool floating in the liquid, e.g. a precipitate of aluminium hydroxide is flocculent.

milky (*adj*) describes a liquid (p.10) with a white precipitate (p.30) which gives the liquid the appearance of milk. The precipitate is very light, e.g. when carbon dioxide is passed into lime water, a light precipitate of calcium carbonate turns the lime water into a milky liquid.

creamy (*adj*) describes a white precipitate which is heavier than the precipitate forming a milky (↑) liquid; the precipitate still floats in the liquid, e.g. silver chloride forms a creamy precipitate.

heavy (*adj*) describes a precipitate (p.30) which sinks to the bottom of the liquid, e.g. barium sulphate forms a heavy precipitate.

miscible (*adj*) describes liquids that can mix in
all proportions (p.76); the result looks like a
single liquid, e.g. water and alcohol can mix
completely and look like one liquid.

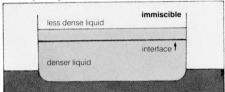

less dense liquid
immiscible
interface
denser liquid

immiscible (*adj*) describes liquids that do not
mix at all, e.g. oil and water form two layers (↓)
of liquid as oil and water are immiscible.

layer (*n*) a spread of material over the surface
(p.16) of another material. A layer can be thick
or thin, e.g. a layer of skin covers an orange; a
sandwich has 3 layers – bread, meat, bread.

film (*n*) a thin layer (↑) of a substance. It can be
a thin layer of a liquid, or a vapour, or a solid; a
thin layer of one liquid on another liquid; a thin
layer of a solid on another solid. For example,
a thin film of oil on water; a thin film of oxide
(p.48) on a metal.

interface (*n*) the meeting place between two
layers of liquid, e.g. oil floats on water, where
the oil and water meet is an interface.

mobile (*adj*) (1) describes a liquid that flows
easily, e.g. kerosene is a mobile liquid. (2)
describes an object that can be moved easily,
e.g. a bicycle is mobile. For liquids, mobile is
the opposite of viscous (↓). **mobility** (*n*).

engine oil

a viscous liquid

viscous

viscous (*adj*) describes a liquid that does not
flow quickly, e.g. engine oil is viscous. Viscous
is the opposite of mobile (↑). **viscosity** (*n*).

volatile (*adj*) describes a liquid that vaporizes
(p.11) easily, e.g. petrol is a highly volatile
liquid. **volatility** (*n*).

viscosity (*n*) the property of a liquid which prevents it from flowing readily, e.g. olive oil has a
high viscosity; water has a very low viscosity.

brown fumes

lead (II) nitrate

heat

lead (II) oxide

chemical change

action/acts on

copper

concentrated nitric acid

copper (II) nitrate

the actions of nitric acid on copper

concentrated nitric acid acts on copper

chemical change (*n*) a change in which new materials or new substances (p.8) are formed, e.g. when chalk is heated, lime and carbon dioxide are formed, this is a chemical change. When sulphur dioxide is passed into a solution of sodium hydroxide, sodium sulphite is formed, this is a chemical change.

chemical property a property which depends on the effect of other substances, e.g. a chemical property of sulphuric acid is that alkalis (p.45) neutralize the acid; a chemical property of water is that it is decomposed (p.65) by sodium.

nature (*n*) the essential (p.226) properties of a material or substance (p.8) from its nature, e.g. the nature of alkalis: they neutralize acids, and have a soapy feel when dissolved in water.

action (*n*) the effect of a substance on another material or substance, or the effect of heat, or an electric current, on a substance (p.8). For example, the action of sulphuric acid on calcium carbonate forms carbon dioxide and calcium sulphate; the action of heat decomposes (p.65) calcium carbonate.

act on to produce a chemical action (↑), e.g. hydrochloric acid acts on calcium carbonate, an action takes place (p.63), carbon dioxide is given off (p.41) and calcium chloride formed.

active (*adj*) (1) describes a substance (p.8) with many chemical properties, or with chemical properties that reacts vigorously, e.g. sulphuric acid is a very active substance, as many actions (↑) occur with other substances, and they are strong actions. (2) describes the part of a mixture (p.54) or molecule (p.77) which has the particular properties exhibited by the whole mixture or molecule, e.g. the hydroxide group (p.185) of an alcohol (p.175) is the active part of the molecule.

inactive (*adj*) describes a substance (p.8) with few chemical properties, e.g. the alkanes (p.172) are inactive substances. *Compare stable (p.74).*

inert (*adj*) describes: (1) a substance (p.8) which has no chemical properties, e.g. krypton is an inert gas; (2) an atmosphere preventing oxidation. *Compare stable (p.74).*

pure (*adj*) describes a substance (p.8) which has no other substances mixed with it, e.g. pure silver has no other substances mixed with it. **purify** (*v*), **purification** (*n*).

impure (*adj*) describes a substance (p.8) which is not pure (↑) as it has other substances mixed with it; these other substances are *impurities*. The amount of the impurities is usually small, e.g. iron usually contains a small amount of carbon, and is therefore impure iron. **impurity** (*n*).

trace (*n*) a very small amount of any substance (p.8) present (p.217) in a mixture or in the earth. Also, a very small amount of an impurity, e.g. a trace of arsenic in sulphur is a very small amount of impurity; a trace element in the earth is an element (p.8) present in very small amounts.

contaminate (*v*) to make a substance impure (↑) with a small amount of an unwanted impurity (↑), e.g. drinking water contaminated with small amounts of dissolved lead; aluminium contaminated with silicon. **contamination** (*n*), **contaminated** (*adj*).

chemical[1] (*adj*) describes a property or action concerned with the making of new materials or substances (p.8).

chemical[2] (*n*) an element (p.8) or compound (p.8) which takes part in a chemical change, e.g. sodium hydroxide is a chemical; sulphuric acid is a chemical.

fine chemical a chemical which is pure (↑). Fine chemicals are used in analysis (p.82) and for other special purposes. Only small quantities are made of some fine chemicals.

pharmaceutical chemical a pure substance used for medicinal purposes.

technical chemical a reasonably pure substance, not as pure as a fine chemical, but purer than a coarse chemical. Technical chemicals are used for most industrial (p.157) purposes. *See heavy chemical (p.171).*

coarse chemical an impure substance used in industry and agriculture. Suitable for the purpose used and cheap to manufacture.

pure

no other substances

silver

impure

small amount of carbon

iron

trace of arsenic

sulphur

trace

pharmaceutical chemical

aspirin tablets

flammable inflammable liquid

flammable/inflammable

non-flammable

non-flammable liquid

corrosive (*adj*) describes any chemical (↑) which attacks the surface (p.16) of solids and of living things, and destroys these surfaces. Examples of corrosive chemicals are strong mineral acids (p.55), such as concentrated sulphuric acid, which destroy skin. *Compare corrosion (p.61)*.

caustic (*adj*) describes a chemical (↑) which attacks and destroys the surface of living things. Strong alkalis (p.45) are described as caustic, e.g. sodium hydroxide is called caustic soda and it burns skin when it comes into contact with it.

mild (*adj*) describes an alkali which is not caustic (↑), but is stronger than a weak alkali. *Mild* describes a condition which is between strong and weak, e.g. mild steel is less strong than hard steel, but harder than iron.

bland (*adj*) describes any chemical (↑) which does not irritate (p.22) nor cause discomfort to people. Bland usually describes food and pharmaceutical chemicals (↑), e.g. olive oil is bland.

passive (*adj*) describes the surface of a substance (p.8) which has been made inactive (p.19). A passive substance may also be inactive because its surface is covered with a thin film (p.18) of oxide (p.48), as on some metals. Usually the surface is made inactive by the attack of a corrosive (↑) chemical, e.g. iron is made passive by the action of concentrated nitric acid. Aluminium is passive because its surface is covered with a thin film of oxide. **passivity** (*n*).

activated (*adj*) describes a substance (p.8) which has been made active (p.19), e.g. activated charcoal has had its power to absorb (p.35) made greater than that of ordinary charcoal.

flammable (*adj*) describes a substance (p.8) which readily bursts into flames.

inflammable (*adj*) another word for flammable (↑).

non-flammable (*adj*) describes a substance which does not burst into flames, and does not burn. The opposite of flammable (↑).

odoriferous (*adj*) describes a material or
substance which produces an odour (p.15),
e.g. onions are odoriferous.

irritate (*v*) to make parts of animals painful; to be
unpleasant to the senses, particularly touch.
For example, wood smoke irritates the eyes and
the nose of a person; the eyes become painful,
and the nose has an unpleasant feeling. **irritant**
(*n*), **irritating** (*adj*).

pungent (*adj*) describes an odour (p.15) which
has a strong effect on the nose or the tongue,
e.g. vinegar has a pungent odour and taste.
The odour is neither pleasant nor unpleasant.

acrid (*adj*) describes an odour (p.15) which is
like the smell of wood smoke. It is an irritating
(↑) odour.

choking (*adj*) describes an odour which is more
irritating (↑) than an acrid (↑) odour; it is felt at
the back of the throat and is very unpleasant,
e.g. the smell of hydrogen chloride gas is a
choking odour.

malodorous (*adj*) describes a material or
substance (p.8) which has a very unpleasant
odour, e.g. bad food is malodorous.

fragrant (*adj*) describes an odour which is
pleasant to the senses, e.g. the odours of fruit
and flowers are often fragrant.

dense (*adj*) (1) describes fumes (p.33) which are
opaque (p.16), e.g. a fire can give off dense
smoke and it is not possible to see through the
smoke. (2) describes any material or substance
(p.8) with a high density (p.12).

tools

saw

chisel

hammer

tank

tap

pinchcocks

rubber tube

rubber tube

device (*n*) an object made for a special purpose, e.g. a telescope is a device for seeing distant objects; a thermostat is a device for keeping a constant (p.106) temperature.

tool (*n*) a device (↑) which is held in the hand and used to help a person who is working with his hands, e.g. a saw, a spanner, a screwdriver.

thermometer

ammeter

instruments

barometer

instrument (*n*) a device (↑) used for measuring, recording (p.39) or detecting (p.225), e.g. a thermometer, a barometer, an ammeter, a spectroscope. The work done with an instrument is more accurate (p.227) than the work done with a tool (↑) and needs more knowledge.

apparatus (*n*) (*apparatus n.pl.*) all the objects, devices (↑), tools (↑) and instruments (↑) used for work in chemistry, or science, e.g. thermometers, beakers, ammeters, thermostats, supports.

laboratory (*n*) a room in which scientific experiments (p.42) are carried out.

tank (*n*) a large container for liquids.

tap (*n*) a device (↑) which controls the flow of a liquid or gas from a pipe.

pinchcock (*n*) a device (↑) which closes a tube (p.29) or a pipe.

stopcock (*n*) a kind of tap (↑) used with glass tubes (p.29).

practical (*adj*) describes work with chemicals on experiments, which is opposite to work with writing and ideas.

theoretical (*adj*) describes work with ideas, usually recorded (p.39) in writing. Theoretical work is the opposite of practical (↑) work.

stopper (*n*) an object which closes a hole in a pipe, a bottle, or a flask (↓), e.g. a glass stopper in a bottle.

cork (*n*) a stopper (↑) made from the bark of a cork tree.

bung (*n*) a stopper (↑) made of rubber, or a large stopper made of wood.

delivery tube a glass tube (p.29) connecting different pieces of apparatus (p.23); it conducts fluids (p.11) from one piece of apparatus to another.

cork

glass stopper

rubber bung

gas jar

delivery tube

pneumatic trough

beehive

connect

rubber tube   glass tube

two glass tubes connected by a rubber tube

connect (*n*) to join together different pieces of apparatus, so that fluids can flow from one part of the apparatus to another.

disconnect (*n*) to separate different pieces of apparatus so that they are not connected (↑). Disconnect is the opposite of connect.

trough (*n*) (1) a flat container for liquids. (2) the hollow between the crests of two waves.

pneumatic trough a flat container used in the collection of gases.

gas-jar (*n*) a tall vessel (↓) for the collection of gases.

beehive (*n*) the stand for a gas-jar which is put in a pneumatic trough (↑).

aspirator (*n*) a large vessel used to supply air or water to apparatus.

eudiometer (*n*) a glass tube (p.29), with a scale, used to measure the volume of gases.

water in   air out

aspirator

eudiometer

receptacle (*n*) a container in which solids, liquids, or gases are put. Any cup, beaker (↓), crucible (p.27), flask (↓) or receiver (p.28) is a receptacle.

vessel (*n*) a receptacle (↑) for containing liquids.

rim (*n*) the outside edge of the opening to a receptacle (↑).

spout (*n*) a pipe leading from a vessel (↑), through which a liquid passes when leaving the vessel.

leak (*v*) to lose fluids from a receptacle or pipe through a small hole in the receptacle or pipe, e.g. water leaks from a hole in a water pipe.

beaker (*n*) an upright vessel for containing liquids and also in which liquids can be heated.

flask (*n*) a round-shaped vessel or bottle with a long narrow neck; it is used to contain liquids for experiments (p.42). Flasks can have round bottoms or flat bottoms.

conical flask a vessel with a narrow neck and a flat bottom; it is used in volumetric analysis (p.82).

Erlenmeyer flask a large conical flask (↑) containing more than 500 cm³ liquid.

U-tube (*n*) a glass tube (p.29) in the shape of the letter U.

Woulfe bottle a glass bottle with two necks; it is used to pass gases through a liquid.

graduation (*n*) a mark on an instrument, or
measuring vessel, which is equally spaced
from other marks to show a scale (↓) for
measurement. **graduated** (*adj*).

scale (*n*) a set of marks, with numbers beside
them, rising from a low value to a high value,
e.g. the scale on a thermometer rising from 0°C
to 100°C. Each mark is a graduation (↑) on the
scale.

measuring cylinder a tall, narrow vessel, with a
scale (↑) for volume; it is used to measure the
volume of liquids. Measuring cylinders can measure volumes of 100 cm³, 1 dm³, etc.

burette (*n*) a long, narrow vessel with a spout and
stopcock; used for measuring the volumes of a
liquid which is allowed to run out of the burette.
Burettes are used in volumetric analysis (p.82).

pipette (*n*) a vessel of a particular shape which
delivers a fixed amount of liquid. Common
sizes of pipettes are: 10 cm³, 20 cm³, 25 cm³,
and 50 cm³.

graduated flask a flask with a very long and very
thin neck. It has a graduation (↑) on the neck
and measures the volume of a liquid very
accurately. Graduated flasks commonly have
sizes of 100 cm³, 250 cm³, 500 cm³ and 1 dm³
They are used in volumetric analysis (p.82).

calibrate (*v*) to make a scale (↑) on a device,
instrument, or other measuring apparatus, so
that it measures accurately, e.g. to pass a
known electric current through an ammeter
(p.123) and to mark the value of the current on
a scale, and then to complete the scale by
marking other graduations. **calibration** (*n*).

separating funnel

funnel

crucible

funnel (*n*) a piece of apparatus with a wide mouth and a thin pipe; it is used to put liquids into bottles, flasks and other vessels.

separating funnel a kind of funnel (↑) used to separate immiscible (p.18) liquids.

porous pot (*n*) a vessel (p.25) with porous (p.15) sides; it is used with gases and liquids, and these fluids (p.11) can pass through the sides.

crucible (*n*) a receptacle (p.25) used for heating solids to a high temperature.

generator (*n*) any piece of apparatus which gives a supply of a required gas, e.g. a generator of hydrogen.

Kipp's apparatus a generator used to give a supply of a gas, especially hydrogen sulphide.

thermostat (*n*) a device used to keep a liquid at a constant (p.106) temperature.

balance[1] (*n*) any piece of apparatus used to weigh solids, and liquids contained in vessels. The *mass* (p.12) of a solid is measured by a beam balance or a lever balance. The *weight* of a solid is measured by a spring balance.

a balance

top loading balance

distillation flask a flask used when distilling
(p.33) liquids; it has a side-arm on its neck.
condenser (*n*) a long glass tube (↓) which has a
larger glass tube around it. Vapour passes
down the inner tube and cold water passes
through the outer tube. The cold water
condenses (p.11) the hot vapour.
receiver (*n*) a vessel (p.25) put at the end of a
condenser (↑) to receive the condensed (p.11)
liquid.
support (*n*) a wooden or iron rod which takes the
weight of a piece of apparatus and prevents
the apparatus from falling. **support** (*v*).
thermometer (*n*) an instrument for measuring
temperature (p.102). Most thermometers use
mercury to measure the temperature on the
Celsius scale. On this scale, water freezes at
0°C and boils at 100°C.
lag (*v*) to put a material (p.8) round a pipe, or
piece of apparatus, to prevent heat escaping.
lagging (*n*).
retort (*n*) a vessel (p.25) with a long neck, used
in distillation (p.33).

retort

steam out **steam trap**

trap passes steam
prevents water
passing

condensed water

steam in

tongs

tubing

tube

stout walled

bore

thin walled

drawing of a beaker

diagram

diagram of
a beaker

trap (*n*) a device (p.23) which allows one fluid
(p.11) to pass through, but prevents another
kind of fluid from passing through, e.g. a steam
trap allows steam to pass through, but prevents
water passing through.

jet (*n*) (1) a very small hole at the end of a pipe.
(2) a stream of fluid which comes out of a jet,
e.g. a jet of water.

blowpipe (*n*) a piece of apparatus like a pipe,
ending in a jet (↑). It can be used to blow a
small, very hot flame onto an object.

spatula (*n*) a tool shaped like a spoon, or which
has a flat surface, used for adding and
removing small quantities of solids.

tongs (*n.pl.*) an article used to hold hot objects.

tube (*n*) a long, hollow article, through which fluids
can flow. Chemical apparatus uses glass and
rubber tubes.

tubing (*n*) different kinds of tubes, either glass or
rubber, may be of different bores (↓).

stout-walled (*adj*) describes tubes (↑) with thick
walls relative to the diameter of the bore (↓).

thin-walled (*adj*) describes tubes (↑) with thin
walls relative to the diameter of the bore (↓).

bore (*n*) the size of the hole in a tube (↑), e.g. a
bore of 5 mm.

ground glass (*n*) glass which has been rubbed
to give a very smooth surface.

quick-fit (*adj*) describes apparatus which uses
ground-glass (↑) stopper (p.24) and joints.

diagram (*n*) a drawing, using lines, to show how
apparatus (p.23) is connected. A diagram is
used because it is simpler to draw a diagram
than to draw a picture.

precipitate

clear solution

precipitation

dissolve (v) (1) to mix a solid (p.10) or a gas (p.11) with a liquid resulting in a solution (p.86), e.g. sugar is dissolved in water, air is dissolved in water. (2) to mix with a liquid, e.g. solids and gases dissolve in water; sugar dissolves in water. The solid or gas is soluble (p.17) if it dissolves.

precipitate (n) a solid (p.10) which appears in a solution when two solutions (p.86) are mixed. It is the result of a chemical reaction (p.62), e.g. when a solution of sulphuric acid is added to a solution of barium chloride, a precipitate appears. Precipitates are described as flocculent, creamy or heavy (see p.17). **precipitate** (v), **precipitated** (adj).

filter (v) to separate an insoluble (p.17) solid from a liquid by pouring through a filter in a funnel. The filter can be filter paper or glass wool. **filter** (n), **filtered** (adj), **filtrate** (n), **filtration** (n).

filtrate (n) the liquid that passes through a filter (↑).

dissolve

water

sugar

sugar dissolves in water

filter paper

residue ———— funnel

filtration

filtrate

sedimentation

suspended particles

sediment

residue (*n*) (1) the solid that does not pass through a filter (↑). A precipitate (↑) is collected as a residue during filtration. (2) the material (p.8) or substances (p.8) left behind after any process, e.g. the residue left in a flask after distillation (p.33); the residue left in an evaporating basin after evaporation (p.11).

suspended (*adj*) describes light particles (p.13) of an insoluble (p.17) solid which float at all levels in a liquid or gas (p.11).

settle (*v*) to fall slowly through a fluid; particles (p.13) of a solid fall slowly to the bottom of a fluid when they settle. For example, insoluble particles fall to the bottom of a liquid; dust settles from the air onto a surface (p.16).

sediment (*n*) solid particles (p.13) that settle (↑) to the bottom of a vessel (p.25) form sediment. **sedimentation** (*n*).

decant (*v*) to pour off a clear (p.16) liquid, leaving any sediment (↑) at the bottom of the vessel (p.25). Before decanting liquid, any suspended (↑) material is allowed to settle as a sediment. *See supernatant (p.90).*

decanting

sediment left in flask

clear liquid

evaporate[2] (*v*) to heat a liquid so that it boils and vapour is given off. This makes the volume of the liquid become less. A liquid can be evaporated to dryness, with all the liquid changed to vapour and dissolved (p.30) solids left as a residue (p.31). **evaporation** (*n*), **evaporated** (*adj*).

evaporating basin a vessel (p.25) used for evaporating (↑) liquids.

concentrate (*v*) to boil away liquid from a solution (p.86) so that the same amount of solid is dissolved (p.30) in less liquid. The concentration (p.81) of the solution is then increased. **concentration** (*n*).

digest (*v*) to make a solid dissolve (p.30) in a liquid by adding the solid to the hot liquid. The hot liquid is usually stirred (↓). To contrast *digest* and *dissolve*: when a solid is *digested* a chemical action (p.19) takes place between the solid and the liquid; when a solid is *dissolved* a physical change (p.13) takes place, and there are no new substances formed. For example, copper (II) oxide is digested in hot sulphuric acid to make copper (II) sulphate. **digestion** (*n*).

stir (*v*) to move a glass rod, or other article, in a circular path in a liquid or powder (p.13) in order to mix the constituents (p.54) together. **stirrer** (*n*).

fuse (*v*) to melt a powdered solid so that it forms one solid mass. **fusible** (*adj*), **fusion** (*n*), **fusibility** (*n*).

calcine (*v*) (1) to heat a solid to a high temperature to drive off volatile (p.18) substances. (2) to heat a metal to a high temperature to form the oxide (p.48) of the metal. **calcined** (*adj*).

ignite (*v*) to set a material or substance on fire so that it burns (p.59). The ignition temperature of a substance is the lowest temperature at which it will catch fire. **ignition** (*n*), **ignited** (*adj*).

deflagration (*n*) the bursting into flame of a substance caused by chemical action, e.g. the deflagration of phosphorus in chlorine. **deflagrate** (*v*).

evaporating basin

liquid

heat

solution concentrated

volume less

evaporation

digestion | stirrer

hot liquid

solid

solid dissolves in liquid

deflagration

fumes of phosphorus trichloride

phosphorus (deflagrates in chlorine)

chlorine gas

pyrolysis (*n*) the decomposition (p.65) of a chemical compound (p.8) by heat. Pyrolysis is usually used for organic (p.55) compounds; the organic compound is decomposed to simpler compounds.

distil (*v*) to change a liquid to a vapour (p.11) by heating and then to condense (p.11) the vapour back to a liquid. **distillation** (*n*).

distillation (*n*) the process (p.157) of distilling a liquid. Distillation is used to separate two or more liquids which have different boiling points. It is also used to purify (p.43) liquids.

vapour rises

liquid boils

water out

vapour condensed

distillation

water in

heat

distilled liquid

distilled (*adj*) describes a liquid which has been purified (p.43) by distillation, e.g. distilled water is very pure water.

bumping (*n*) the forming of very large bubbles (p.40) when a liquid boils. As a bubble rises, the vessel containing the liquid jumps up and down; this event is called boiling by bumping. To prevent bumping, small pieces of porous pot (p.27) are put in the liquid.

sublimation

sublimate

cool surface

vapour rises

solid

heat

sublime (*v*) to change a solid to a vapour (p.11) by heating, and then to cool the vapour so that it changes directly back to a solid, e.g. to sublime sulphur. **sublimation** (*n*), **sublimed** (*adj*).

sublimation (*n*) the process (p.157) of subliming a solid. Sublimation is used to purify solids. Not many solids sublime.

sublimate (*n*) a solid substance that has been formed (p.41) by sublimation.

fumes (*n.pl.*) (1) small particles (p.13) in the air, looking like smoke, are fumes. (2) vapour given off by an acid and combining with water vapour in the air to give the appearance of smoke. (3) any visible (p.42) vapour, especially a vapour which irritates (p.22) the nose and eyes. **fumes** (*v*).

separation (*n*) the way in which substances (p.8) are separated from each other, especially liquids. Immiscible (p.18) liquids are separated by using a separating funnel (p.27). Miscible (p.18) liquids are separated by distillation (p.33). Different methods are used for the separation of solid mixtures (p.54). **separate** (*v*), **separable** (*adj*).

extraction (*n*) (1) the process (p.157) of taking one substance from a mixture (p.54) of substances, e.g. the extraction of a solid by using a solvent (p.86) such as the extraction of iodine from its solution in water by using tetrachloromethane as a solvent. (2) the process of obtaining an element (p.8) from the earth; in a few cases the element can be extracted directly; in most cases an ore (p.154) is taken from the earth and the element extracted from the ore. **extract** (*v*).

dialysis (*n*) a process (p.157) for the separation (↑) of a colloid (p.98) from a crystalloid (p.91). A mixture of a colloid and a crystalloid is put in a receptacle (p.25) made from a membrane (p.99), usually parchment or cellophane. Water is passed round the membrane; the crystalloid passes through the membrane and is taken away by the water. The colloid remains in the membrane. **dialyze** (*v*). ·

separation

separation of two immiscible liquids

water in

crystalloid + water out

water

crystalloid + colloid

crystalloid passes into water

colloid remains

dialysis

diffusion

hydrogen
effuses out

porous plug
glass tube

hydrogen gas
(gas under
pressure)

water

effusion

diffusion (*n*) (1) the process (p.157) of one gas spreading through another gas, e.g. petrol vapour spreads through the air in a room by diffusion. (2) when two miscible (p.18) liquids are put together, they spread through each other by diffusion. (3) when a soluble solid is put in a liquid, the solid dissolves and spreads through the liquid by diffusion. (4) gases pass through a porous (p.15) membrane (p.99) by diffusion if the pressure on each side of the membrane is the same. **diffuse** (*v*).

effusion (*n*) the process (p.157) of a gas passing through a porous (p.15) membrane (p.99) or through a small hole when the gas goes from a higher pressure to a lower pressure. Compare diffusion (↑) where the pressure is the same on both sides of the membrane or hole. **effuse** (*v*).

absorption (*n*) (1) the process (p.157) of a solid taking in a liquid or a gas, e.g. the absorption of a gas by charcoal, the absorption of water by a gel (p.100). (2) the process of a liquid taking in a gas, e.g. sodium hydroxide solution absorbs carbon dioxide from the air.

chromatography (*n*) a method of separating a mixture of solutes, by using a solvent and a separating medium which can be paper or an acid material such as silica or alumina, or an inert (p.19) support coated with a solvent. In gas chromatography, volatile constituents (p.54) are made to flow through a column packed with an inert support coated with a suitable solvent by an inert gas.

cork
glass support
paper strip
glass vessel

paper chromatography

mixture in solution
second solvent

paper chromatography a solid mixture is dissolved in a suitable solvent and drops of this solution are placed on marks on a strip of filter paper. The paper strip is put in a corked container with one end in a second solvent, *see diagram*. The second solvent moves up the paper strip and takes with it the constituents (p.54) of the mixture. Different constituents rise to different heights.

development (*n*) if the constituents of a mixture used in paper chromatography (↑) are colourless, then the paper is sprayed with a dilute solution of a reagent which forms a coloured derivative (p.200) of a constituent. This action, to produce different coloured derivatives, is called *development*. **develop** (*v*).

solvent front the greatest height reached by a solvent in paper chromatography (↑). The constituents of a mixture never reach this height.

chromatogram (*n*) a paper strip, or a column, in chromatography (↑) with the individual constituents marked by coloured spots, using development (↑) if necessary, after separation by a suitable solvent.

chromatogram

paper strip
solvent front
location of constituents
original mark

column chromatography

column chromatography a glass tube is packed
with an inert (p.19) solid, usually aluminium
oxide, forming a column. A mixture is placed
on the top of the column. A suitable solvent is
added above the mixture, *see diagram*. The
solvent flows slowly down the column taking
the constituents (p.54) of the mixture with it.
Different constituents travel different distances.
By continually adding solvent, all the
constituents will eventually pass out of the
column and can be separately collected.

eluent (*n*) the solvent used to separate a mixture
in column chromatography (↑). This method of
separation is called *elution*.

location (*n*) the process of finding the positions
of constituents (p.54) in a chromatogram (↑).
For colourless constituents, development (↑)
or ultra-violet light is used to produce colours.
locate (*v*).

grind (*v*) to change large lumps of a solid into a powder by rubbing in a pestle and mortar (↓). **ground** (*adj*).

pestle (*n*) an article with a handle and a rounded glass or stone end, used to grind (↑) substances in a mortar (↓).

mortar (*n*) a receptacle (p.25) for holding substances to be ground by a pestle.

triturate (*v*) to mix by grinding (↑). Two or more solids, or a solid and a liquid can be triturated. **trituration** (*n*).

treat (*v*) to add a reagent (p.63) or any chemical to a material or substance to cause a chemical change; e.g. to treat iron with concentrated nitric acid to make the iron passive; to treat cotton with sodium hydroxide solution to give the cotton a shiny surface (p.16). **treatment** (*n*).

acidify (*v*) to add acid (p.45) to a solution (p.86) so that there is excess (p.230) acid present. **acidified** (*adj*).

litmus paper   in acids   in alkalis
indicator

indicator (*n*) a chemical substance (p.8) that shows whether a solution is acidic (p.45), alkaline (p.45) or neutral (p.45). **indicate** (*v*).

indicate (*v*) to show clearly, or to make a sign.

pH (*abbr*) a way of describing how acid or how alkaline a solution is.

pH value (*n*) a number on a scale of 1 to 14 showing the strength or concentration of an acid or alkali. A value of pH = 1 indicates (↑) a strong acid. A value of pH = 14 indicates a strong alkali. A value of pH = 7 indicates a neutral solution.

pH scale

| 0 | 2 | 4 | 6 | 7 | 8 | 10 | 12 | 14 |

strong   weak   neutral   weak   strong

acid   alkali

first reading

acid

alkali

titration

titration (*n*) the process (p.157) of letting a solution (p.86) flow from a burette (p.26) into another solution held in a conical flask (p.25) until a chemical reaction (p.62) is complete. A common titration is between an acid (p.45) and an alkali (p.45) using an indicator (p.38). **titrate** (*v*).

end point the sign that a titration (↑) is complete as the chemical reaction has been completed. In an acid and alkali titration the end point is shown by a change in the colour of an indicator.

titre (*n*) the volume of a solution from a burette needed to reach the end point in a titration (↑).

reading (*n*) the value on a scale of an instrument (p.23) or a piece of apparatus (p.23) which is taken as a measurement, e.g. the reading of the level of the solution in a burette from the scale of volume on the burette; mercury in a thermometer shows the temperature on the thermometer scale, this is the reading on the thermometer.

record (*v*) to write down a reading (↑) or an observed (p.42) effect. **record** (*n*).

result (*n*) a change which is measured or observed (p.42). To contrast *result* and *effect*: when iron is exposed to air and water it rusts (p.61) forming iron oxide; rusting is the effect, and iron oxide is the result. **result in** (*v*).

tabulate (*v*) to write down results (↑) in a table. **tabulation** (*n*).

table of results	cm³	cm³	cm³
2nd reading	24.1	46.7	26.3
1st reading	1.4	24.1	3.6
titre	22.7	22.6	22.7

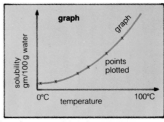

graph (*n*) a line drawn to show the relation between two changing quantities, e.g. pressure and volume of a gas, solubility (p.87) of a substance and temperature. **graphical** (*adj*).

plot (*v*) to make a graph (↑) by putting marks for results (↑) and connecting the marks by a line.

evolve (*v*) to form bubbles (↓) of gas and give off
(p.41) the gas. To contrast *form*, *give off*, and
evolve gases: *form* and *give off* can be used
with both physical and chemical changes, *form*
describes a weaker effect than *give off*; *evolve*
is used only with chemical changes and
describes a stronger effect than *give off* and
the gas or vapour evolved can be collected.
Evolution of gases are described as *steady*,
brisk and *rapid* as the quantity of gas evolved
is increased. **evolution** (*n*).

effervescence

effervesce (*v*) to evolve (↑) a gas rapidly with the
formation of many bubbles (↓) at a liquid
surface (p.16), e.g. when a dilute acid is added
to a carbonate, the mixture effervesces.
effervescence (*n*).

bubble[1] (*n*) a small quantity of a gas or vapour
with liquid around it. The liquid can be a thin
film (p.18) round the bubble or the liquid can
be in a vessel with bubbles in the liquid.

bubble[2] (*v*) to cause bubbles (↑) of gas to go
through a liquid.

form (*v*) to cause to come into being, e.g. when two solutions (p.86) are mixed, a precipitate (p.30) is formed; kerosene forms a vapour; hydrogen is formed when calcium metal reacts with water. **formation** (*n*).

give off of an object, living thing, or chemical action, to cause a gas, vapour (p.11), or odour (p.15) to come from itself. For example, when iron is added to dilute sulphuric acid, hydrogen (a gas) is given off by the reaction; when water is boiled, steam is given off by the water; a flower gives off an odour. *See liberate (p.69).*

generate (*v*) to produce (p.62) large quantities of a gas for a particular purpose. *See generator (p.27).* **generator** (*n*).

displaced air out

dense gas in

collect

collecting a dense gas

hydrogen gas in

pass over

passing hydrogen gas over a heated solid

copper (II) oxide

heat

out

gas

collect

collecting a gas

gas given off

generator

water

solid

heat

collect (*v*) (1) to obtain a gas in a vessel (p.25), usually a gas-jar (p.24). (2) to obtain a distilled liquid in a receiver (p.28). (3) to obtain a specimen (p.43) of crystals (p.91).

pass over to cause a gas to flow over a solid, e.g. hydrogen is passed over heated copper (II) oxide.

investigate (*v*) to study carefully by means of experiments (↓), tests (↓) and recorded (p.39) facts, e.g. to investigate the properties (p.9) of sulphuric acid is to find and record the properties by making a careful study of them. **investigation** (*n*).

experiment (*n*) to work with instruments and apparatus in a practical (p.23) investigation (↑) on the behaviour and nature of substances.

visible (*adj*) describes anything that can be seen. *See perceptible* (↓).

observation (*n*) the use of the senses for a particular purpose, e.g. when a chemical reaction (p.62) takes place, a person makes observations on any changes of state, colour, or odour, and on any new substances formed. Observations are recorded as results (p.39). **observe** (*v*), **observable** (*adj*).

perceptible (*adj*) describes any changes, usually physical changes, which can be observed (↑) by the senses, although such changes are small. Observable changes can be noticed more easily than perceptible changes. Changes can be *readily, barely* or *hardly* perceptible, showing increasing difficulty of perception. **perception** (*n*), **perceive** (*v*).

test (*n*) the use of a chemical reagent (p.63) to identify (p.225) a substance, a metal, a cation or anion (p.125) or any group of substances. The test gives a result and from the result an inference (↓) is made. Some tests give a definite (p.226) result so the identification is made without doubt. **test** (*v*).

confirmatory (*adj*) describes a test which makes an identification without doubt. **confirm** (*v*), **confirmation** (*n*).

test paper (*n*) a piece of paper with a reagent (p.63) on it; the paper changes colour when testing particular substances and thus identifies (p.225) these substances, e.g. litmus paper.

demonstrate (*v*) to show clearly by practical examples, e.g. to demonstrate the properties of chlorine gas by showing in an experiment (↑) the action of various substances on the gas. **demonstration** (*n*).

test

silver nitrate solution

solution under test

white precipitate

inference

chloride carbonate sulphate

nitric acid

precipitate insoluble

a test for chloride

confirms chloride

copper (II) oxide

sulphuric acid

heat

digest the oxide in the acid

solution of copper (II) sulphate

filter off impurities

filtrate

concentrate the filtrate

heat

leave to form crystals

a preparation of copper (II) sulphate

preparation

inference (*n*) the use of results (p.39) to decide on the identification (p.225) of a substance. An inference is only an opinion, a confirmatory (↑) test has to be made to be sure. **infer** (*v*).

conclusion (*n*) (1) the use of results (p.39) to decide on the identification (p.225) of a substance without doubt, e.g. inferences (↑) from tests together with a confirmatory (↑) test allow a conclusion to be made. (2) the use of results to decide on the relation (p.232) between observations (↑), e.g. a set of results on pressure and volume allows a conclusion to be drawn on the relation between the pressure and the volume of a gas. **conclude** (*v*).

technique (*n*) an accepted way of carrying out (p.157) a process (p.157) which needs practical (p.23) skill and a knowledge of chemistry, e.g. if a person knows the technique of distillation (p.33), he can connect suitable apparatus, heat to the correct temperature and collect a pure distillate.

isolate (*v*) to obtain a pure substance from a mixture (p.54) of substances. The pure substance can be a compound (p.8) or an element (p.8), e.g. to isolate bromine from sea water.

preparation (*n*) (1) a substance made in a laboratory (p.23) for a particular purpose, e.g. a student makes crystals of copper (II) sulphate, the substance is a preparation. (2) to make ready the apparatus and any solutions (p.86) needed in an experiment (↑). **prepare** (*v*).

specimen (*n*) (1) a small quantity of a material or substance (p.8) which is isolated (↑) from a mixture. (2) a quantity of a substance which is used as an example of that substance, i.e. the properties of the specimen are the properties of all quantities of the substance. For example, tests carried out (p.157) on a specimen of copper demonstrate (↑) the properties of the specimen and thus the properties of copper.

purification (*n*) a process (p.157) or processes carried out (p.157) to remove impurities (p.20) from a substance, e.g. the purification of silver by which impurities such as lead are removed (p.215). **purify** (*v*).

trivial name	quick lime	blue vitriol	oil of vitriol
traditional name	calcium oxide	cupric sulphate	sulphuric acid
systematic name	calcium oxide	copper (II) sulphate	sulphuric acid

trivial name	green vitriol	common salt	—
traditional name	ferrous sulphate	sodium chloride	potassium permanganate
systematic name	iron (II) sulphate	sodium chloride	potassium manganate (VII)

nomenclature (n) a way of naming chemical substances (p.8).

chemical nomenclature

trivial name a nomenclature (↑) for chemical substances which was used before chemistry was studied properly. Examples of trivial names are alum, blue vitriol, lime, and chalk.

traditional name a nomenclature (↑) for chemical substances which shows their chemical composition (p.82); this method was used before system-·atic names (↓) were used. Some traditional names are still used, and some are the same as systematic names. Examples of traditional names are cupric sulphate and lead nitrate.

systematic name the modern nomenclature (↑) for chemical substances. For inorganic (p.55) compounds, the oxidation number (p.78) of the metal is given, and the acid radical (↓) or anion (p.125) is described with the oxidation number for its important element (p.8), e.g. copper (II) sulphate, iron (III) sulphate, potassium manganate (VII). For organic (p.55) compounds, the systematic name is taken from a suitable alkane (p.172), e.g. *ethanoic* acid is taken from *ethane*; ethanoic acid is the systematic name of acetic acid (traditional name).

binary compound a compound (p.8) formed from the chemical combination (p.64) of two elements (p.8). The systematic name (↑) for such compounds ends in -*ide*, e.g. lead (II) oxide, calcium carbide, phosphorus trichloride.

radical (*n*) a group of atoms, which are part of a molecule (p.77) or form an ion (p.123); the radical often remains unchanged throughout a chemical reaction (p.62) or a series of reactions. The atoms of a radical which forms an ion are held together by covalent bonds (p 136). Some radicals are also functional groups (p.185). Examples of radicals are the sulphate ion, the nitrate ion, the manganate (VII) ion and the ammonium ion.

acid radical a radical (↑) combined (p.64) with hydrogen in an acid (↓).

acid (*n*) a substance (p.8) which contains (p.55) hydrogen which can be replaced (p.68) by a metal, or by a base (p.46). An acid is a covalent (p.136) substance, which when dissolved in water produces hydrogen ions (p.123) in the solution. The strength of an acid is measured by its pH value (p.38). **acidify** (*v*), **acidic** (*adj*).

alkali (*n*) a soluble base (p.46). The solution of an alkali contains hydroxide (p.132) ions; it reacts with an acid (↑) to produce a salt (p.46) and water only. **alkaline** (*adj*), **alkalinity** (*n*).

acidic (*adj*) (1) describes a substance having the nature (p.19) of an acid (↑). (2) describes a solution (p.86) containing an acid. (3) describes a compound (p.8) which forms an acid when dissolved in water, e.g. sulphur dioxide is an acidic oxide.

alkaline (*adj*) describes a solution (p.86) with the properties of an alkali, e.g. sodium hydroxide solution is alkaline.

neutral (*adj*) describes a substance or a solution which is neither acidic (↑) nor basic (p.46). A neutral solution has a pH value (p.38) of 7. **neutralize** (*v*).

neutralization

acid **+** alkali → salt + water

base (*n*) a substance which reacts (p.62) with an acid (p.45) to produce a salt (↓) and water only. Bases are usually the oxides (p.48) or hydroxides (p.48) of metals. Many bases are insoluble. **basic** (*adj*), **basicity** (*n*).

basic (*adj*) describes a substance having the nature of a base (↑), e.g. copper (II) oxide is a basic oxide.

amphoteric (*adj*) describes a substance which has both acidic (p.45) and basic (↑) properties, e.g. aluminium hydroxide reacts with **acids** to form a salt and water and it also reacts with **alkalis** to form a salt and water.

salt (*n*) a compound made by replacing some or all of the hydrogen of an acid (p.45) by a metal. A base or an alkali or a metal reacts with an acid to replace the hydrogen by a metal or an ion (p.123) such as the ammonium ion. Salts in solution usually form ions (p.123). Examples of salts are copper (II) sulphate (in which copper is combined (p.64) with the sulphate radical (p.45) as it has replaced the hydrogen of sulphuric acid), iron (III) chloride and lead (II) nitrate.

basicity (*n*) the number of hydrogen atoms which can be replaced (p.68) in one molecule of an acid; also the number of hydrogen ions (p.123) formed from one molecule of an acid. For example, hydrochloric acid HCl has one atom of hydrogen that can be replaced in one molecule of the acid, so it has a basicity of one; sulphuric acid H_2SO_4 has two replaceable hydrogen atoms, therefore it has a basicity of two.

monobasic (*adj*) describes an acid with a basicity (↑) of one.

dibasic (*adj*) describes an acid with a basicity (↑) of two.

tribasic (*adj*) describes an acid with a basicity (↑) of three.

normal salt a salt formed when all the hydrogen atoms in an acid (p.45) molecule have been replaced by a metal or an anion (p.125) of a metal, e.g. sodium chloride, copper (II) sulphate, sodium carbonate.

copper (II) sulphate
salt + water

acid salt a salt formed when not all the hydrogen atoms in an acid (p.45) molecule have been replaced by a metal or a cation (p.125), e.g. potassium hydrogensulphate, calcium hydrogencarbonate. Only dibasic (↑) and tribasic (↑) acids can form acid salts.

basic salt a salt formed when not all the base (↑) reacts with the acid (p.45) in a reaction. The base combined (p.64) with the normal salt (↑) to form an insoluble basic salt, e.g. basic lead carbonate, formed from lead (II) carbonate and lead (II) oxide.

aluminium sulphate

potassium sulphate

double salt

crystals of alum (aluminium potassium sulphate)

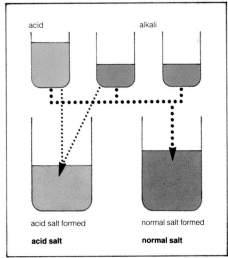

acid

alkali

acid salt formed

normal salt formed

acid salt

normal salt

double salt a compound (p.8) of two normal salts (↑) which forms crystals (p.91), e.g. aluminium potassium sulphate, a compound of aluminium sulphate and potassium sulphate, known as alum.

complex salt a compound (p.8) containing a complex ion (p.132); either the metal or the ion (p.123) of a normal salt, or both, can form a complex ion. An example is tetraammine copper (II) sulphate (cuprammonium sulphate); *see also hexacyanoferrate (p.53).*

hydrated (*adj*) describes a salt (p.46) with water chemically combined (p.64) with the compound in its crystals (p.91). The combined water is called *water of crystallization*, e.g. copper (II) sulphate .5H$_2$O, in which five molecules of water are combined with one of the salt in its crystals.

water of crystallization see *hydrated* (↑).

anhydrous (*adj*) describes a salt with no water of crystallization (↑), e.g. sodium chloride crystals have no water of crystallization.

anhydride (*n*) (1) a compound (p.8) formed by removing (p.215) the elements (p.8) of water from a substance, but no water of crystallization (↑), e.g. ethanoic acid (acetic acid) can have the elements of water removed to form ethanoic anhydride. (2) a compound which, when added to water, forms a new chemical compound; the anhydride is named after this new chemical compound, e.g. ethanoic anhydride, when added to water, forms ethanoic acid.

oxide (*n*) a binary compound (p.8) formed by the combination of an element (p.8) with oxygen. The element can be a metal or a non-metal (p.116), e.g. lead (II) oxide, carbon dioxide.

hydroxide (*n*) a compound (p.8) formed by the reaction of a basic (p.46) oxide with water and having the radical (p.45) −OH. A soluble hydroxide forms hydroxyl (p.132) ions in water, e.g. sodium hydroxide NaOH, calcium hydroxide Ca(OH)$_2$ (formed by the action of water on calcium oxide, a basic oxide).

peroxide (*n*) an oxide which reacts (p.62) with cold dilute sulphuric acid to produce hydrogen peroxide. Peroxides in water form the ion (p.123) $(O-O)^{2-}$. Example: sodium peroxide.

higher oxide (*n*) an oxide with an element showing a higher oxidation number (p.78) than usual, e.g. manganese (VII) oxide is a higher oxide with manganese (IV) oxide the usual oxide.

oxides

hydroxide sodium hydroxide

peroxide   sodium peroxide

higher oxide

manganese (VII) oxide

H—C≡C—H   acetylene (ethyne)

$Ca^{2+}(C≡C)^{2-}$ calcium carbide

$(C≡C)^{2-}$ carbide ion (acetylide)

carbonate ion

carbonic acid

carbide (*n*) a binary compound (p.44) containing a metal combined (p.64) with carbon. There are several different kinds of carbides, including the acetylides (↓).

acetylide (*n*) a carbide (↑) which reacts with water to form acetylene (p.174), C_2H_2. The metals of Group I and Group II of the periodic system (p.119) form acetylides. The compounds are ionic (p.123); an acetylide ion with an electrovalency of 2 is formed. Examples of acetylides are Na_2C_2 (sodium acetylide) and CaC_2 (calcium carbide or calcium acetylide).

carbonate (*n*) a compound (p.8) of a metal with a carbonate ion (p.123). Carbonates are salts (p.46) of carbonic acid, a very weak acid formed by dissolving carbon dioxide gas in water. The ion, *see diagram*, is trigonal planar (p.83) in structure (p.82), with three coordinate bonds (p.136).

hydrogen carbonate (*n*) an acid salt (p.47) of carbonic acid; only one hydrogen atom is replaced (p.68) by a metal, e.g. sodium hydrogencarbonate, $NaHCO_3$.

bicarbonate (*n*) the traditional name (p.44) for a hydrogencarbonate.

sodium hydrogen carbonate

chloride (*n*) a compound (p.8) of an element (p.8)
and chlorine. Metals form ionic (p.123)
chlorides; non-metals form covalent (p.136)
chlorides, e.g. Na$^+$Cl$^-$ (ionic chloride); CCl$_4$
(covalent chloride). *See chloro group (p.187)*.
Chlorides are salts (p.46) of hydrochloric acid.
bromide (*n*) a compound (p.8) of an element (p.8)
and bromine. Metals usually form ionic (p.123)
bromides; non-metals form covalent (p.136)
bromides. *See bromo group (p.187)*. Bromides
are salts (p.46) of hydrobromic acid.

**some
sodium
halides**

sodium chloride

sodium bromide

sodium iodide

iodide (*n*) a compound (p.8) of an element (p.8)
with iodine. Metals usually form ionic (p.123)
iodides; non-metals form covalent (p.136)
iodides. *See iodo group (p.187)*. Iodides are
salts (p.46) of hydroiodic acid.
halide (*n*) a compound (p.8) which is a chloride,
a bromide, an iodide, or a fluoride. It is formed
from an element (p.8) and a halogen (p.117).

sulphuric acid
(covalent)

sulphurous acid
(covalent)

sulphate ion
(tetrahedral)

sulphide (*n*) a binary compound (p.44) of an element (p.8) with sulphur. Metals form ionic (p.123) sulphides, most of which are insoluble. Non-metals form covalent (p.136) sulphides. Sulphides are salts (p.46) of the weak acid, hydrogen sulphide, H_2S.

sulphate (*n*) a compound (p.8) of a metal with a sulphate ion (p.123). Sulphates are salts (p.46) of sulphuric acid, a strong acid which is dibasic, forming normal and acid salts. The sulphate ion, *see diagram*, it tetrahedral (p.83) in structure (p.82) with four coordinate bonds (p.136).

hydrogen sulphate (*n*) an acid salt of sulphuric acid, i.e. an acid sulphate (↑). For example, sodium hydrogensulphate $NaHSO_4$ which forms the ion HSO_4^-.

bisulphate (*n*) traditional name (p.44) for hydrogen sulphate.

sulphite (*n*) a compound (p.8) of a metal with a sulphite ion (p.123). Sulphites are salts (p.46) of sulphurous acid, a dibasic acid forming normal and acid salts. The sulphite ion, *see diagram*, is trigonal pyramidal (p.84) in structure (p.82) with three coordinate bonds (p.136) and a lone pair (p.133) of electrons.

hydrogen sulphite (*n*) an acid salt of sulphurous acid, i.e. an acid sulphite, e.g. sodium hydrogensulphite $NaHSO_3$ with the ion HSO_3^-

bisulphite (*n*) traditional name for hydrogen-sulphite (↑).

thiosulphate (*n*) a compound (p.8) of a metal with a thiosulphate ion (p.123). Thiosulphates are the salts (p.46) of thiosulphuric acid $H_2S_2O_3$. The thiosulphate ion, is tetrahedral (p.83) in structure (p.82) with four coordinate bonds (p.136).

sulphite ion   (trigonal pyramidal)

thiosulphate ion (tetrahedral)

nitride (*n*) a binary compound (p.44) of a metal with nitrogen. The common nitrides are formed with the metals of groups I, II and III of the periodic system (p.119). These nitrides react (p.62) with water to form ammonia and the hydroxide of the metal:

nitrate (*n*) a compound (p.8) of a metal with a nitrate ion (p.123). Nitrates are salts (p.46) of nitric acid, a strong acid which is monobasic (p.46). The nitrate ion, *see diagram*, is trigonal planar (p.83) in structure (p.82) with three coordinate bonds (p.136).

nitrite (*n*) a compound (p.8) of a metal with a nitrite ion (p.123). Nitrites are salts (p.46) of nitrous acid, a monobasic acid (p.45). The nitrite ion, *see diagram*, is non-linear (p.83) with two coordinate bonds (p.136).

chromate (VI) (*n*) a compound (p.8) of a metal with the chromate (VI) ion (p.123). Chromic acid, the acid of the salts (p.46), cannot be isolated (p.43). Although the acid is dibasic, only the normal salts (p.46) can be made. The chromate (VI) ion, *see diagram*, is tetrahedral (p.83) in structure (p.82) with four coordinate bonds (p.136).

chromate (*n*) traditional name (p.44) for chromate (VI).

dichromate (VI) (*n*) a compound of a metal with the dichromate (VI) ion (p.123). Dichromic acid, the acid of the salts (p.46), cannot be isolated (p.43). Although the acid is dibasic, only the normal salts (p.46) can be made. The dichromate (VI) ion, *see diagram*, has the structure of two tetrahedra with six coordinate bonds (p.136) and two covalent bonds (p.136).

dichromate (*n*) traditional name (p.44) for dichromate (VI).

nitrate ion   trigonal planar

nitric acid

nitrite ion   (non-linear)

nitrous acid

chromate (VI) ion tetrahedral

manganate (VI).ion
tetrahedral

manganate VII ion
tetrahedral

hexacyanoferrate (II) ion

manganate (VI) (*n*) a compound (p.8) of a metal with the manganate (VI) ion (p.123). The acid of these salts (p.46) cannot be isolated (p.43); no acid salts are formed, only normal salts, although the acid is dibasic (p.46). The manganate (VI) ion, *see diagram*, is tetrahedral (p.83) in structure (p.82) with four coordinate bonds (p.136). The manganese atom has an oxidation number (p.78) of 6. An example of a salt is potassium manganate (VI) K_2MnO_4.

manganate (*n*) traditional name (p.44) for manganate (VI).

manganate (VII) (*n*) a compound (p.8) of a metal with the manganate (VII) ion (p.123). The acid of these salts (p.46) cannot be isolated (p.43); it is monobasic (p.46). The manganate (VII) ion, *see diagram*, is tetrahedral (p.83) in structure (p.82) with four coordinate bonds (p.136). The manganese atom has an oxidation number of 7. An example of a salt is potassium manganate (VII) $KMnO_4$.

permanganate (*n*) traditional name (p.44) for manganate (VII).

hexacyanoferrate (II) (*n*) a compound (p.8) of a metal and the hexacyanoferrate (II) ion (p.123). There is no acid for these salts (p.46). The hexacyanoferrate (II) ion, *see diagram*, is octahedral (p.83) in structure (p.82) with six coordinate bonds (p.136). The iron atom has an oxidation number of 2. An example of a salt is potassium hexacyanoferrate (II) $K_4Fe(CN)_6$. The hexacyanoferrate ion is a complex ion (p.132).

ferrocyanide (*n*) the traditional name (p.44) for hexacyanoferrate (II).

hexacyanoferrate (III) (*n*) a compound (p.8) of a metal and the hexacyanoferrate (III) ion. There is no acid for these salts (p.46). The hexacyanoferrate (III) ion has the same structure as the hexacyanoferrate (II) ion (↑), but the iron atom has an oxidation number (p.78) of 3; it is a complex ion. An example of a salt is potassium hexacyanoferrate (III) $K_3Fe(CN)_6$.

ferricyanide (*n*) the traditional name (p.44) for hexacyanoferrate (III).

mixture (*n*) different substances (p.8), put together, form a mixture. The substances can be
elements (p.8), compounds (p.8) or materials
(p.8), e.g. a mixture of charcoal (the element,
carbon), sulphur (an element) and potassium
nitrate (a compound) forms a mixture called
gunpowder. A particular substance can be
separated (p.34) from a mixture. Mixtures can
be solid, liquid, or gaseous (p.11). **mix** (*v*).

constituent (*n*) (1) a member of a mixture (↑),
e.g. a mixture of sulphur and copper has two
constituents; charcoal, sulphur and potassium
nitrate mixed together form gunpowder,
sulphur is one of the constituents of
gunpowder. Each constituent keeps its own
properties in a mixture. A liquid mixture can be
a true solution (p.86), a colloidal solution (p.98)
or a suspension (p.86). (2) a part of a
compound (p.8), e.g. a molecule of sulphur
dioxide consists of (p.55) one atom of sulphur
and two atoms of oxygen, it has three
constituent atoms. Constituents of a
compound are elements (p.8), radicals (p.45),
functional groups (p.185) and ions (p.123).
constitute (*v*), **constitution** (*n*).

ingredient (*n*) a substance (p.8) needed for a
mixture (↑) before it is put in the mixture, e.g.
charcoal, sulphur and potassium nitrate are
ingredients of gunpowder before they are
mixed to form gunpowder; after mixing they are
constituents (↑).

homogenous (*adj*) describes a material or
substance (p.8) which is the same throughout
in its properties (p.9) and composition (p.82).
Homogenous also describes a chemical
reaction (p.62) with all substances in the same
state of matter (p.9), e.g. all in the gaseous
state, or all in the liquid state. **homogeneity**
(*n*), **homogenize** (*v*).

heterogenous (*adj*) describes a material,
substance (p.8), or chemical reaction (p.62)
which is not the same throughout in its
properties (p.9), composition (p.82), or state of
matter (p.9). It is the opposite of homogenous
(↑). **heterogeneity** (*n*).

ingredients of
gunpowder   **mixture**
potassium
nitrate

sulphur

charcoal

mix together

gunpowder

a mixture

sulphur is a
constituent

mixture

mixtures of
iron and sulphur
varied proportions

compound of
iron and sulphur
constant proportions

metal
1 part lead

metal
2 parts tin

melt together

solder
an alloy
(a mixture
of metals)

alloy

alloy (*n*) a mixture (p.54) of two or more metals, or of a metal and a non-metal. The mixture is homogenous (p.54) and a new material is formed. An alloy makes a particular metal more useful, e.g. silver is soft, but a silver alloy is hard and more useful than pure (p.20) silver. **alloy** (*v*).

brass (*n*) an alloy (↑) of 60 – 90% copper with zinc; other metals may be added also. Brass is harder than pure copper, and is more useful for a number of purposes.

amalgam (*n*) an alloy (↑) of mercury with other metals. An amalgam is usually soft and may even be liquid. **amalgamate** (*v*).

consist of to have as constituents (↑) when all the constituents are named, e.g. gunpowder *consists of* charcoal, sulphur and potassium nitrate; brass *consists of* copper and zinc.

contain (*v*) to have as constituents (↑) when some, or only one, of the constituents are named but not all, e.g. gunpowder contains sulphur and charcoal; gunpowder contains sulphur.

organic (*adj*) describes substances (p.8) (and the study of these substances in chemistry) which are compounds (p.8) of carbon, but not the oxides or the carbonates of carbon. All substances in living things are organic, but many organic substances are not found in living things. *See hydrocarbon (p.172) and carbohydrate (p.205)*.

inorganic (*adj*) describes substances (p.8) (and the study of these substances in chemistry) which are not organic (↑). Such substances are generally obtained from minerals (p.154).

mineral acid (*n*) an acid obtained from a mineral (p.154) by chemical processes (p.157). The three most important mineral acids are hydrochloric, sulphuric and nitric acids.

mineral acid

three mineral acids
(inorganic acids)

hydrochloric
acid

HCl

sulphuric
acid

H_2SO_4

nitric
acid

HNO_3

air (*n*) a mixture (p.54) of gases which forms the atmosphere and is the cause of atmospheric pressure (p.102). Air contains (p.55) about 20% oxygen, 79% nitrogen, 1% noble gases (↓) and 0.3% carbon dioxide; this is the composition (p.82) of dry air. In addition air always contains water vapour (p.11).

diluent (*n*) a substance added to a solution or to a mixture of solids or gases to reduce (p.219) the concentration of the solution or to reduce the proportion (p.76) of one of the constituents (p.54) of the mixture, e.g. water can be added to a concentrated (p.88) alkali as a diluent; nitrogen in the air is a diluent for oxygen.

noble (*n*) (1) *noble* gases of the air (↑) describes the gases helium, neon, argon, krypton and xenon. The gases are generally considered inert (p.19). (2) *noble* metals describes gold, platinum and other metals which do not react with the usual mineral acids.

pollution (*n*) undesirable substances (p.8) in the air, water or earth; the surroundings (p.103) become unhealthy or impure, e.g. the pollution of the air by smoke.

water (*n*) a liquid substance formed by the combination (p.64) of hydrogen and oxygen (formula: H_2O).

water vapour pressure all water on the Earth's surface evaporates (p.11) and so air contains water vapour. The vapour exerts (p.106) a pressure, which is part of atmospheric pressure (p.102).

saturated water vapour pressure The water vapour pressure (↑) when the air is saturated (p.87) with water vapour. It is the highest water vapour pressure at a particular temperature. Saturated water vapour pressure increases with a rise in temperature.

water cycle a cycle (p.64) in which water evaporates (p.11) from the Earth's surface and forms clouds, the clouds break to form rain, the rain passes through the Earth to rivers, lakes and finally the sea. Water evaporates from the rivers, lakes and sea. Respiration (p.61) also helps in the cycle.

atmosphere

air forms the atmosphere

air

water cycle

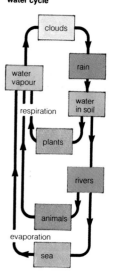

hard water water which does not form a lather
(↓) easily with soap. It contains (p.55) salts of
calcium and magnesium which form insoluble
salts with the soap. **hardness** (*n*).

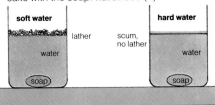

soft water water which forms a lather (↓) easily
with soap. It does not contain salts of calcium
and magnesium, or else such salts have been
removed (p.215). **soften** (*v*).

lather (*n*) a large quantity of very small bubbles
(p.40) formed when soap is mixed with water.
lather (*v*).

temporary hardness a hardness of water (↑)
which can be removed (p.215) by boiling the
water. It is caused by calcium hydrogencarbonate being dissolved in the water.

permanent hardness a hardness of water (↑)
which cannot be removed (p.215) by boiling. It
is removed by (a) adding sodium carbonate,
(b) by detergents (p.171), or (c) by zeolite (↓).
It is caused by dissolved salts of calcium and
magnesium, such as the sulphate.

water softening any process by which hard water
(↑), with either temporary or permanent
hardness (↑), is changed to soft water (↑).

zeolite process a process (p.157) using minerals
(p.154), called zeolites, to soften (↑) water. A
zeolite contains sodium ions (p.123) which can
be replaced by other metal ions. In water
softening, the zeolite removes (p.215) the
calcium and magnesium ions from hard water
and replaces (p.68) them with sodium ions.

contamination (*n*) the presence in water, food,
or any other substance or material (p.8) of
causative agents of disease, e.g. the presence
of viruses, bacteria, protozoa, etc. which cause
disease.

water softening

water softener

hard water in

combustion (*n*) any chemical reaction (p.62) in which heat, and usually light, is produced. Combustion commonly is the burning of organic (p.55) substances during which oxygen from the air (p.56) is used to form carbon dioxide and water vapour. **combustible** (*adj*).

rapid combustion combustion (↑) in which heat is produced (p.62) at a high temperature, usually with flames, by the combination (p.64) of substances with oxygen.

slow combustion combustion (↑) in which heat is produced (p.62) at a low temperature, without flames, by the combination (p.64) of substances with oxygen.

explosion (*n*) a chemical reaction which takes place very quickly, forming large volumes of gases and releasing (p.69) energy (p.135). The energy causes a rise in temperature. An explosion produces (p.62) sound and often light. The increase in pressure caused by the large volume of gas causes destruction in the surroundings. **explode** (*v*), **explosive** (*adj*).

hot gases give out light

combustion of hydrogen

chemical reaction between hydrogen and oxygen

oxygen in air

hydrogen gas

unburnt gas

flame

flame (*n*) combustion (↑) of gases produces (p.62) a flame. The hot gases give out light.

spontaneous combustion combustion (↑) which takes place without any apparent (p.223) cause, e.g. the spontaneous combustion of phosphorus in oxygen.

incombustible (*adj*) describes any substance which does not undergo (p.213) combustion (↑).

glow

smoulder

burn (v) to undergo (p.213) combustion (↑) producing (p.62) heat, flame and sometimes smoke, e.g. wood burns with a smoky flame; to burn coal to produce heat.

glow (v) to give out light because of a chemical reaction (p.62) or because a substance (p.8) is heated strongly, e.g. a piece of wood glows when there are no flames; a piece of iron glows when heated strongly, **glow** (n).

smoulder (v) to give out smoke because of a chemical reaction (p.62) taking place at a low temperature without a flame.

warm (v) to supply only enough heat to make substances react (p.62) or a process (p.157) to take place. To raise the temperature of a liquid or a solid so that it can still be touched without discomfort. For example, to warm a solution so that crystals dissolve. **warm** (adj).

heat (v) to supply enough heat to raise the temperature so that a substance is hot enough for a chemical reaction (p.62) to take place or a process (p.157) to take place. The temperature is too high for the substance to be touched with comfort. For example, to heat zinc in a stream of oxygen. **hot** (adj), **heat** (n), **heater** (n).

char (v) to change an organic (p.55) substance to carbon by heating it or by using a strong dehydrating (p.66) agent. For example, to char wood by heating it, forming charcoal; to char sugar by adding concentrated sulphuric acid.

water-bath (n) a vessel (p.25) containing water, which is used to heat (↑) apparatus (p.23). This way of heating does not allow the temperature to rise above 100°C.

water-bath

substances heated to react

temperature 100°C or less

steam

water

water bath

heat

heating by water-bath

red-hot (*adj*) describes a substance, usually a metal, heated to such a high temperature that it glows (p.59) red.

incandescent (*adj*) describes a substance heated to such a high temperature that it gives out white light. The temperature of an incandescent substance is higher than the temperature of a red-hot (↑) substance. **incandescence** (*n*).

light¹ (*n*) a form of energy whose effect on the eye causes the sense of seeing. White light is composed (p.82) of all colours. Different colours can cause different chemical effects. Some substances, e.g. silver chloride, are decomposed by light falling on them. *See photochemical (p.65).*

burst into to appear suddenly, as flames do, e.g. when a lighted match is put near petrol, the petrol bursts into flames. *Burst into* is used when a lot of flames appear; with a small flame, the substance *catches fire.*

light² (*v*) to put a flame to a supply of a gas, so that the gas catches fire, i.e. flames appear, e.g. to light a supply of hydrogen from a jet.

extinguish (*v*) to cause a flame to stop burning.

photosynthesis

carbon cycle

respiration (*n*) a process (p.157) in plants and animals by which oxygen reacts (p.62) with carbohydrates (p.205) to form carbon dioxide and water vapour, with the release (p.69) of energy (p.135). It is a kind of slow combustion (p.58). **respire** (*v*).

photosynthesis (*n*) a process (p.157) in green plants by which carbon dioxide from the air and water from the soil is used to make carbohydrates (p.205). Energy (p.135) from the sun is used in the chemical reaction (p.62). **photosynthesize** (*v*), **photosynthetic** (*adj*).

carbon cycle a cycle (p.64) of chemical reactions in which green plants convert (p.73) carbon dioxide to carbohydrates by photosynthesis (↑). The carbohydrates are used by plants and animals in respiration (↑) and the carbon dioxide is returned to the air (p.56).

corrode (*v*) to form a substance on the surface of a metal. The metal reacts (p.62) with particular gases in the air to form the substance; in many cases an oxide is formed. When a metal corrodes, small holes appear on the surface and the strength of the metal becomes less. Oxygen, carbon dioxide, sulphur dioxide or hydrogen sulphide in the air cause a metal to corrode. **corrosion** (*n*), **corrosive** (*adj*).

rust (*n*) corrosion (↑) of iron, forming a red dust which is the rust on the surface of the iron. Rust is an oxide of iron. **rust** (*v*), **rusty** (*adj*).

tarnish (*v*) to corrode if the metal has a shiny surface, e.g. silver tarnishes by the forming of a layer of silver sulphide on the surface; silver sulphide is a corrosion, or tarnish. **tarnish** (*n*).

chemical reaction a process (p.157) in which
new substances (p.8) are formed, i.e. a
chemical change takes place. Energy, usually
heat, is needed to make the chemical reaction
take place, or the chemical reaction produces
($\downarrow$) heat.

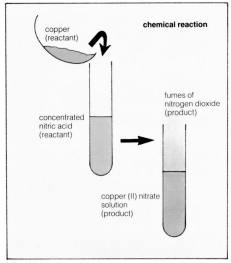

reaction (*n*) in chemistry, a process (p.157) in
which two substances (p.8) have an effect on
each other and new substances are produced
($\downarrow$). **react** (*v*).
react (*v*) to behave in such a way that a chemical
reaction ($\uparrow$) takes place, e.g. sodium metal
reacts with water, a chemical reaction takes
place and sodium hydroxide and hydrogen are
formed (sodium and water react).
reactant (*n*) a substance which takes part in a
chemical reaction ($\uparrow$), e.g. sodium and water
are reactants when these substances react.
product (*n*) a new substance formed from a
chemical reaction ($\uparrow$), e.g. hydrogen and
sodium hydroxide are the products of the
reaction between sodium metal and water.
produce (*v*).

reagent (*n*) a substance which causes a chemical reaction (↑) to take place. The mineral acids (p.55) are common reagents as they have known effects on many inorganic (p.55) substances.

hydrogen peroxide

oxidising agent

agent (*n*) a substance, or a form of energy, which is used to produce a named effect, e.g. an oxidizing agent causes oxidation (p.70); light causes a photochemical (p.65) effect and so is an agent for the effect.

take place to come into being, when describing an event, i.e. an event takes place. When an event *happens*, it is not expected, when an event *takes place*, it is planned. *See occur[1] (↓)*.

tin (II) chloride

reducing agent

occur[1] (*v*) to come into being at a certain time when describing an event, e.g. an event occurs, it is known when it will occur, although a person does not control it. To contrast *take place* and *occur*: a chemical reaction (↑) *takes place* because a person controls it, it cannot *occur* because it has not happened before and it is under the control of a person. A birthday occurs once a year, it does not take place.

reversible reaction | hydrogen + oxide of iron | hydrogen | steam + iron
steam | | |
red-hot iron — heat | | oxide of iron — heat |

reversible reaction a reaction (p.62) in which
reactants (p.62) form products (p.62) and the
products can then react to form the reactants.
For example, steam and iron form hydrogen
and an oxide of iron. The oxide of iron and
hydrogen react to form steam and iron. The
reaction is written thus:
    steam + iron $\rightleftharpoons$ oxide of iron + hydrogen
irreversible reaction a reaction (p.62) in which
the products will not react and the reaction
stops when the products are formed.
chain reaction (*n*) a reaction (p.62) in which the
first reaction between molecules (p.77) or atoms
(p.110) forms products which then react with
further molecules or atoms so that each reaction
becomes stronger until the reaction is explosive.
cycle (*n*) a set of events which is repeated time
after time. Each complete set is one cycle, and
the cycle is repeated. For example, a carbon
atom starts in a molecule of carbon dioxide in
the air; it is used in photosynthesis (p.61) in a
plant and becomes part of a carbohydrate
(p.205) molecule; the carbohydrate is eaten by
an animal and digested; the digested product
takes part in respiration (p.61) and the carbon
atom becomes part of a carbon dioxide molecule, and is breathed out; the carbon atom is
once again in a carbon dioxide molecule in the
air, the cycle has been completed, and starts
all over again. **cyclic** (*adj*).
combination (*n*) the joining of elements (p.8) by
chemical bonds (p.133) to form compounds (p.8).
The joining of compounds by addition (p.188) to
form new substances, e.g. the combination of
iron and oxygen forms iron (III) oxide; the
combination of zinc and chlorine forms zinc
chloride; the addition of bromine to ethene
(p.174) is a combination of the two compounds.

chain reaction

● atoms
● neutrons

decomposition

copper (II) carbonate

heat   carbon dioxide

decomposes

copper (II) oxide

decomposition (*n*) the changing of pure substances into simpler compounds. The process (p.157) is irreversible. Decomposition can be caused by temperature, light, electric current or micro-organisms, e.g. the decomposition of copper (II) carbonate when heated into copper (II) oxide and carbon dioxide; the decomposition of silver bromide by light into silver and bromine; the decomposition of copper (II) sulphate by an electric current. **decompose** (*v*).

dissociation (*n*) the separation of an ionic (p.123) or a covalent (p.136) compound (p.8) into simpler compounds; the process (p.157) is reversible, i.e. the simpler compounds can unite again. For example, sodium chloride, in water, undergoes (p.213) dissociation into ions (the ions can unite to form sodium chloride again); the dissociation of ammonium chloride into ammonia and hydrogen chloride (ammonia and hydrogen chloride can combine to form ammonium chloride). **dissociate** (*v*).

thermal (*adj*) caused by heat.

photochemical (*adj*) describes a chemical reaction caused by light. *See electrochemical (p.128).*

thermal decomposition decomposition (↑) caused by heating a compound.

thermal dissociation

ammonium chloride dissociates   combination

hot   fumes of ammonia hydrogen chloride   cold

ammonium chloride

heat

thermal dissociation dissociation (↑) caused by heating a compound, e.g. the thermal dissociation of ammonium chloride.

disintegrate (*v*) to break into small pieces because of physical or chemical action, e.g. when hit hard a mineral (p.154) disintegrates into small pieces. **disintegration** (*n*).

hydrolysis[1] (*n*) decomposition (p.65) caused by the chemical action of water. The salts of weak acids or weak bases undergo (p.213) hydrolysis, e.g. iron (III) chloride in water decomposes into iron (III) hydroxide and hydrochloric acid, as iron (III) hydroxide is a weak base.

dehydrate (*v*) (1) to take away water from a substance, to make it dry, e.g. to take water from ethanol (ethyl alcohol) to make it as dry as possible. (2) to take away the elements (p.8) of water from a compound by a chemical action, e.g. concentrated sulphuric acid removes water from ethanol to form ethene:

dehydrating ethanol

dehydrate

$$C_2H_5OH \rightarrow C_2H_4 + (H_2O \text{ taken away})$$

Dehydrate is the opposite process of *hydrate* (p.90). **dehydration** (*n*).

desiccate (*v*) to take away all traces of water, i.e. a stronger action than dehydrate (↑). **desiccation** (*n*), **desiccator** (*n*), **desiccant** (*n*).

desiccant (*n*) a substance which will desiccate (↑) compounds (p.8).

desiccation

deliquesce (*v*) to absorb (p.35) water from the air by crystals (p.91) so that the crystals dissolve (p.30) and form a concentrated solution, e.g. sodium hydroxide flakes deliquesce in air and form a solution.
deliquescent (*adj*), **deliquescence** (*n*).

effloresce (*v*) to lose water of crystallization (p.91) from a crystal to the air, e.g. sodium carbonate crystals effloresce in air and form a powder. $Na_2CO_3.10H_2O \rightarrow Na_2CO_3.H_2O$. **efflorescent** (*adj*), **efflorescence** (*n*).

hygroscopic (*adj*) describes crystalline (p.15) and amorphous (p.15) substances which take water from the air and become damp, e.g. sodium chloride is hygroscopic and becomes damp when left in air. **hygroscopicity** (*n*).

neutralization (*n*) the reaction (p.62) between an acid (p.45) and a base (p.46) or alkali, during which they destroy each other's properties and form a salt and water. The solution of the salt is neutral, i.e. neither acidic nor alkaline. For example, the neutralization of hydrochloric acid by sodium hydroxide solution. **neutralize** (*v*), **neutral** (*adj*).

displacement (*n*) a chemical reaction (p.62) in which one element takes the place of another element which is in a compound. For example, iron put into a solution of copper (II) sulphate causes the displacement of copper. The reaction is: $Fe + CuSO_4 \rightarrow Cu + FeSO_4$ Copper metal appears in the solution as the iron disappears. **displace** (*v*).

replace (*v*) (1) to put one thing in place of another because the thing replaced is no longer wanted or no longer of use. (2) to put one atom (p.110), an ion or a functional group (p.185) in place of another atom, ion or group. Usually hydrogen in an acid is replaced by a metal, when the metal acts on (p.19) the acid to form a salt and hydrogen. **replacement** (*n*), **replaceable** (*adj*).

replaceable (*adj*) describes a hydrogen atom in an acid which can be replaced, directly or indirectly, by a metal atom.

base exchange a chemical reaction (p.62) in which two inorganic radicals (p.45) replace (↑) each other, e.g. as in the reaction:
calcium chloride + sodium carbonate $\rightarrow$ calcium carbonate + sodium chloride
In this example the chloride and carbonate radicals exchange with each other.

copper (II) sulphate solution

iron

iron displaces copper

iron (II) sulphate solution

copper

displacement

base exchange

sodium carbonate

calcium chloride

calcium carbonate

sodium chloride

release (v) to set free something held physically or to set free energy. Usually a gas in a liquid is released, e.g. when a finely divided (p.13) solid is put into a solution of carbon dioxide, the carbon dioxide gas is released into the air; when a gas tap is opened, gas is released into a burner. **release** (n).

liberate

hydrogen

hydrogen liberated

hydrochloric acid

zinc   zinc replaces hydrogen

liberate (v) to set free a gas by breaking chemical bonds (p.133), e.g. when zinc metal is put in hydrochloric acid, hydrogen is liberated, as bonds in the acid have been broken. To compare *give off* (p.41) with *liberate*: hydrogen is *given off* when zinc is put in hydrochloric acid, this is a practical observation (p.42) as no information is given about the reaction; hydrogen is *liberated* when zinc is put in hydrochloric acid gives information about the chemical reaction. **liberation** (n).

affinity (n) a measure of the ability of one element or compound to react (p.62) with another compound. The greater the affinity, the stronger is the reaction, e.g. sodium hydroxide has an affinity for carbon dioxide; concentrated sulphuric acid has a great affinity for water, the action is violent.

moderate (adj) describes a reaction midway between weak and strong. Many reactions are moderate and so are not usually described as such; only weak, strong, violent (↓) and explosive (p.204) reactions are so described.

violent (adj) describes a reaction which is stronger than strong; the strength of the reaction can break apparatus unless care is taken. **violence** (n).

oxidation (*n*) (1) the addition of oxygen to an element or compound. (2) the removal (p.215) of hydrogen from a compound. (3) the removal of electrons (p.110) from an atom (p.110) or ion (p.123). (4) an increase in the oxidation number (p.78) of an element. For example: (1) calcium + oxygen forms calcium oxide; (2) hydrogen chloride is oxidized to chlorine by the removal of hydrogen; (3) a Cu atom is oxidized to Cu^{2+}, or a Fe^{2+} ion is oxidized to Fe^{3+}; (4) Mn(VI) is oxidized to Mn(VII). **oxidize** (*v*), **oxidizing** (*adj*), **oxidant** (*n*).

oxidation and reduction

hydrogen chlorine

oxidation

reduction

hydrogen chloride   chlorine

Fe^{3+} Fe^{2+} Fe

←oxidation←   ←oxidation←

→reduction→   →reduction→

electron

oxidation and reduction

copper

oxidation

reduction

copper (II) oxide

oxygen   copper

reduction[1] (*n*) (1) the removal (p.215) of oxygen from a compound. (2) the addition of hydrogen to an element or compound. (3) the addition of electrons (p.110) to an atom (p.110) or ion (p.123). (4) a decrease (p.219) in the oxidation number (p.78) of an element. For example: (1) the reduction of zinc oxide to zinc; (2) the reduction of chlorine to hydrogen chloride; (3) the reduction of Cu^{2+} to Cu or of Fe^{3+} to Fe^{2+}; (4) Mn(VII) is reduced to Mn(IV). **reduce** (*v*), **reducing** (*adj*).

redox process if substance A oxidizes (↑) substance B, then substance B reduces (↑) substance A. Oxidation and reduction always occur together, so such a reaction is called a redox process, e.g. hydrogen reduces copper (II) oxide to copper, but copper (II) oxide oxidizes hydrogen to water (hydrogen oxide), the reaction is a redox process.

redox reaction — copper reduced, iron oxidized same reaction (displacement of copper by iron)

Fe + Cu²⁺

oxidation — reduction

Fe²⁺ + Cu

autoxidation (*n*) an oxidation by oxygen in the air at room temperature, e.g. the oxidation of iron in the air to form rust (p.61).

disproportionation (*n*) a process (p.157) in which a substance undergoes (p.213) oxidation (p.70) and reduction (p.70) at the same time, e.g. Cu^+ undergoes disproportionation into Cu and Cu^{2+} One atom of Cu^+ is oxidized to Cu^{2+} and one atom is reduced to Cu.

$$Cu_2^+SO_4^{2-} \rightarrow Cu + Cu^{2+}SO_4^{2-}$$

oxidizing agent a substance which oxidizes (↑) elements or compounds, e.g. oxygen is an oxidizing agent, it oxidizes iron to iron (III) oxide; potassium manganate (VII) solution is an oxidizing agent, it oxidizes iron (II) sulphate to iron (III) sulphate.

oxidant (*n*) another name for oxidizing agent (↑).

reducing agent a substance which reduces (↑) compounds, e.g. hydrogen is a reducing agent, it reduces copper (II) oxide to copper; tin (II) chloride solution is a reducing agent, it reduces iron (III) sulphate to iron (II) sulphate.

catalyst (*n*) a substance which increases the rate of chemical reaction (p.62) but itself remains chemically unchanged, e.g. platinum is a catalyst for the reaction between sulphur dioxide and oxygen, it increases the rate of reaction. Many finely divided (p.13) metals act as catalysts. **catalyze** (*v*), **catalysis** (*n*).

catalysis (*n*) a process in which a catalyst (↑) increases the rate of reaction (p.149).

negative catalyst a substance which decreases (p.219) the rate of reaction (p.149); it reacts with and destroys the catalyst.

inhibitor (*n*) a substance which slows down a chemical reaction (p.62); it may be a negative catalyst (↑) or a retarder (↓).

retarder (*n*) a substance used to decrease the rate of reaction (p.149) by physical and other means, e.g. the addition of a colloid (p.98) to slow down an ionic (p.123) reaction.

promoter (*n*) a substance used to increase the action of a catalyst (↑), e.g. finely divided (p.13) iron catalyzes the reaction between nitrogen and hydrogen to form ammonia, the addition of iron oxide increases the effect of the catalyst, that is, iron oxide is a promoter.

autocatalysis (*n*) a process (p.157) which takes place when one of the products (p.62) of a chemical reaction (p.62) acts as a catalyst (↑) for the reaction, e.g. potassium manganate (VII) in acid solution oxidizes ethanedioic (oxalic) acid, the Mn^{2+} ions, a product of the reaction, catalyze the reaction so autocatalysis takes place.

enzymatic (*adj*) describes any effect caused by enzymes. An enzyme is a catalyst (↑) produced by living things and which takes part in chemical changes in living things. **enzyme** (*n*).

poison (*n*) a substance which prevents a catalyst (↑) from acting, e.g. arsenic is a poison which prevents platinum from acting as a catalyst. Substances which are poisonous for catalysts are also poisonous for living things as they prevent enzymatic (↑) reactions in living things.

catalysis

oxygen

sulphur dioxide

platinum catalyst

sulphur trioxide

convert

transmute

transmutation of
magnesium by neutrons

convert (v) (1) to change, by chemical reaction
(p.62), an element or a compound to another
compound that is wanted for a particular purpose. (2) to change the physical nature (p.19)
of a material (p.8), a substance, or apparatus.
For example to convert nitrogen to ammonia;
to convert nitrogen to ammonia; to convert iron
to steel; to convert water to steam. *See transform (p.144)*. **conversion** (n), **converter** (n).

transmute (v) to change one element (p.8) into
another, using a radioactive (p.138) change or
using bombardment (p 143) of subatomic
particles (p.110). **transmutation** (n).

decolorize (v) to take away colour so that a
coloured substance becomes colourless (p.15),
e.g. iodine solution is brown, but when iodine
is converted to iodide ions, by adding potassium iodide, the iodine solution is decolorized
and becomes colourless. **decolorization** (n).

bleach (v) to take colour from a coloured
substance, usually a pigment (p.162), leaving
the substance white, e.g. to bleach cotton with
chlorine, the natural yellow colour of cotton is
changed to white; to bleach paper using sulphur
dioxide. Chlorine and sulphur dioxide are the
common bleaches. **bleach** (n), **bleaching** (n).

detonate (*v*) to initiate (↓) an explosion (p.58), using an electric spark, or similar agent, called a **detonator** (*n*).

initiate (*v*) to cause a process (p.157) to start when the agent of the cause takes no further part in the process, e.g. chlorine and methane do not react under room conditions, but heating the mixture decomposes (p.65) chlorine molecules (p.77) to chlorine atoms; these atoms react with methane and afterwards the reaction continues without heating, i.e. heat initiates the reaction. **initiation** (*n*).

decrepitate (*v*) to burst with small explosive (p.58) sounds, when crystals (p.91) are heated, e.g. lead (II) nitrate crystals decrepitate when heated and many small explosions can be heard. **decrepitation** (*n*).

ice

thermolabile

stable solid
below 0°C

ice
above 0°C

ice becomes a
thermolabile substance

water

stable

between 0°–100°C
stable liquid

stable (*adj*) describes a substance (p.8) which is not readily changed by heat, chemical reagents (p.63), light, or other forms of energy, e.g. calcium carbonate is stable at all temperatures except very high ones. *Compare inert (p.19), inactive (p.19).* **stability** (*n*), **stabilizer** (*n*), **stabilize** (*v*).

stabile (*adj*) another name for stable (↑).

labile (*adj*) describes a substance (p.8) that readily changes, either physically or chemically, i.e. it decomposes readily. Under normal conditions (p.103), a labile substance is stable (↑), but a change in conditions can easily produce a change in the substance.

unstable (*adj*) describes a substance which tends (p.216) to decompose without apparent (p.223) causes, e.g. nitrogen triiodide is unstable when dry, an insect touching the dry powder can cause it to explode.

thermostable (*adj*) describes a substance which is stable (↑) when heated.

thermolabile substance

copper (II) carbonate

copper (II) oxide

moderate heat

thermolabile

thermolabile (*adj*) describes a substance which loses its nature (p.19) or decomposes when heated.

spontaneous (*adj*) describes an event which takes place without any apparent (p.223) cause, e.g. an unstable (↑) compound can explode spontaneously. **spontaneity** (*n*).

instantaneous (*adj*) describes an event which lasts such a short time that the length of time cannot be measured, e.g. when a match is struck the appearance of the flame is instantaneous. **instant** (*n*).

nascent (*adj*) describes an element at the moment of its coming into being, e.g. when zinc acts on hydrochloric acid, the hydrogen evolved is nascent at the moment of formation.

nascent

nascent hydrogen

hydrochloric acid

zinc

theory (*n*) a description of the causes and effects of natural events and processes (p.157). The theory must rest on experimental observation (p.42). Certain parts of the theory may have to be assumed (p.222) as observation on these parts is not possible, e.g. an atom (p.110) cannot be observed, its properties are assumed. A theory can be used to describe, explain, classify (p.120) and predict (p.85) natural events or processes. **theoretical** (*adj*).

atomic theory the theory that all solids, liquids and gases are made up of atoms (p.110).

molecules of copper (II) oxide

molecules of carbon dioxide

○ copper
● oxygen
● carbon
● nitrogen
○ hydrogen

molecules of ammonia **Dalton's atomic theory**

Dalton's atomic theory all elements (p.8) are made up of small particles (p.110) called atoms. An atom cannot be made, destroyed or divided. Atoms of the same element are alike. Compounds (p.8) are made by atoms combining chemically to form a molecule. A molecule has a small, whole number of atoms. All molecules of a compound are alike. Chemical change takes place when small, whole numbers of atoms combine or are separated.

law of constant proportions all pure specimens (p.43) of the same chemical compound (p.8) contain the same elements (p.8) combined in the same proportions (↓) by mass.

proportion (*n*) if several ratios (p.79) are equal, then their figures are in proportion, e.g. 3/4 = 6/8, so 3 is to 4 as 6 is to 8, the figures are in proportion. **proportional** (*adj*).

combining weight the weight of an element or radical (p.45) which will combine (p.64) with, or displace (p.68), 1 g of hydrogen or 8 g of oxygen. This name is not much used in modern chemistry.

molecule (n) the smallest particle (p.110) of an element or compound (p.8) that can exist (p.213) by itself. A molecule generally consists of a group of atoms (p.110) combined by covalent (p.136) bonds. Ionic (p.123) compounds do not form molecules. **molecular** (adj).

symbol (n) a letter or sign used to show a chemical element, a quantity, a mathematical operation, or a piece of apparatus, e.g. Na is used for sodium, Cl for chlorine, V for volume, p for pressure. See diagram.

symbols in chemistry			
Na	sodium	H	hydrogen
K	potassium	C	carbon
Cu	copper	O	oxygen
Fe	iron	N	nitrogen
Mn	manganese	S	sulphur
Ba	barium	Cl	chlorine
Zn	zinc	I	iodine
Mg	magnesium	P	phosphorus
Ca	calcium	Ag	silver

symbols for quantities			
t	time	p	pressure
m	mass	T	temperature
l	length	c	concentration
V	volume	n	mole fraction
ρ	density	L	Avogadro constant
R	molar gas constant	F	Faraday constant

formula (*n*) (*formulae n.pl.*) (1) a chemical formula shows the number of atoms (p.110) of each element in a molecule or the ions (p.123) in a compound. These are other kinds of formulae; *see structural formula (p.181)*. (2) a physical formula shows the relation (p.232) between different quantities (p.81), e.g. pV = constant; this is a formula showing the relation between the pressure and the volume of a mass of gas.

formula weight this name has been replaced by *relative formula mass* ($\downarrow$).

relative formula mass the mass of one mole (p.80) of molecules (p.77), or ions (p.123) of a compound (p.8). It is calculated from the formula ($\uparrow$) of a compound and the *relative atomic masses* (p.113) of the elements in the compound.

equation (*n*) the formulae ($\uparrow$) of compounds or elements are arranged in an equation to show the reactants (p.62) and products (p.62) of a chemical reaction. For example:

$$2KNO_3 \xrightarrow{heat} 2KNO_2 + O_2$$

This shows (1) two molecules of potassium nitrate decompose (p.65) on heating to form two molecules of potassium nitrite and one molecule of oxygen OR (2) two moles (p.80) of potassium nitrate, when heated, decompose to form two moles of potassium nitrite and one mole of oxygen. Ionic (p.123) equations show reactions between ions. For example:

$$Ba^{2+} + SO_4^{2-} \rightarrow Ba^{2+}\ SO_4^{2-}$$

oxidation number a number showing the oxidation state (p.135) of a metal in a compound (p.8) with the number written in roman numerals, e.g. lead (II) nitrate, with lead having an oxidation number of II and an oxidation state of $+2$; in potassium manganate (VI), maganese in the acid radical (p.45) has an oxidation number of VI and an oxidation state of $+6$.

balance[2] (*v*) (1) to keep an object still, under equal and opposite forces. (2) to have equal numbers of atoms of each element on the opposite sides of an equation, e.g. $2KNO_3 = 2KNO_2 + O_2$ is balanced because there are two potassium atoms, two nitrogen atoms and six oxygen atoms on each side of the equation. **balanced** (*adj*).

---

chemical formulae

CuO
copper oxide

CO$_2$
carbon dioxide

NH$_3$
ammonia

BaSO$_4$
barium sulphate

---

physical formulae

$pV = nRT$

$p = \dfrac{m}{V}$

calculation of density
mass of metal = 296.25 g
volume of metal = 37.5 cm³
density of metal = 296.25 g / 37.5 cm³
= 7.9 gcm⁻³

calculation

calculation
26.7
× 6
160.2
+ 31.3
191.5
− 46.7
144.8
a calculation with numbers

calculate (v) to get a result (p.39) by using arithmetical processes on numbers or the values of quantities (p.81), e.g. to calculate density (p.12) from measurements on the mass and volume of a specimen (p.43). **calculation** (n).

error (n) a mistake in a calculation (↑) or in a statement (p.222). The error can be caused by incorrect calculation, readings (p.39), inferences etc.

exact (adj) describes a calculation (↑) or statement (p.222) without errors (↑).

approximate (adj) describes a measurement which is sufficiently correct to be used in a calculation, e.g. the relative atomic mass of oxygen is 15.99, an approximate value is 16, good enough for most calculations. **approximation** (n).

ratio (n) the ratio of two numbers, or of two quantities (p.81), is obtained by dividing one by the other and simplifying the fraction, e.g. the ratio of 28 to 40 is 28/40 = 7/10, i.e. the ratio is 7:10.

multiple (adj) describes (1) an object made up of two or more like parts. (2) a number which consists of a smaller number multiplied a number of times, e.g. 55 and 44 are multiples of 11.

fixed (adj) describes a quantity (p.81) that has been made unchanging instead of allowing it to change or to be changed, e.g. a metal cylinder contains a fixed volume of gas; the volume has been made unchanging because the volume of the cylinder does not change. **fix** (v).

arbitrary (adj) describes something that has been decided because it is the most suitable or the easiest thing to do and has no relation to any theory (p.76), e.g. the choice of the metre as the standard of length was an arbitrary choice.

mean (adj) describes a value of a quantity (p.81) or a number which is equally far from a high and a low value. If there are two numbers x and y, then $(x + y)/2$ is the arithmetic mean, and $\sqrt{x\,y}$ is the geometric mean, e.g. $6\frac{1}{2}$ is the arithmetic mean of 9 and 4; 6 is the geometric mean of 9 and 4. **mean** (n).

average (adj) describes a value (or a number) obtained by adding together all the values and dividing by the number of values, e.g. the average of 14, 16 and 18 is (14 + 16 + 18) ÷ 3 = 16.

1 mole hydrogen 6.02 × 10²³ molecules	1 mole oxygen 6.02 × 10²³ molecules
22.4 dm³ molar volume (at s.t.p.)	22.4 dm³ molar volume (at s.t.p.)
1 mole carbon dioxide 6.02 × 10²³ molecules	1 mole ammonia 6.02 × 10²³ molecules
22.4 dm³ molar volume (at s.t.p.)	22.4 dm³ molar volume (at s.t.p.)

mole

mole (*n*) the standard (p.229) for measurement of an amount (↓) of substance. One mole of a substance is that amount of substance which contains the same number of elementary particles (p.13) as there are atoms in 0.012 kg (12 grams) of carbon-12. (The isotope (p.114) of carbon with a mass number (p.113) of 12). The particles can be ions (p.123), atoms (p.110), molecules (p.77), electrons (p.110) or any other named particle. If the particles are not named, they are assumed (p.222) to be atoms in elements, molecules in covalent (p.136) compounds or ions in an electrovalent (p.134) compound. **molar** (*adj*).

mole fraction the amount of a substance expressed as a fraction of a mole (↑), e.g. 0.2 mole is a mole fraction.

molar volume the volume of one mole (↑) of a substance in a named state of matter (p.9). The molar volume of any gas at s.t.p. (p.102) is always 22.4 dm³. The symbol V_m is used for molar volume.

Avogadro constant the number of particles in 1 mole (↑). The approximate value is 6.02 × 10²³ The symbol for the Avogadro constant is *L*.

amount (*n*) the amount of a substance is proportional (p.76) to the number of elementary particles (p.13) it contains. The amount of substance is a physical quantity (↓).

quantity (*n*) (1) any measurement of materials, substances (p.8) or energy (p.135) is a quantity. Examples of quantities are mass, length, time, temperature, amount of matter, atomic number, mass number, wavelength, concentration, heat, density. *Compare quality (p.15)*. (2) a measurement of materials or substances that does not give a value, e.g. a quantity of lime; this could be measured in kilograms or moles (↑). **quantitative** (*adj*).

concentration (*n*) the amount (↑) of solute (p.86) dissolved in a solution (p.86). Concentration can be stated as: (a) grams of solute in 1 dm³ solution; (b) moles of solute in 1 dm³ solution; (c) as a percentage. For example, 80 g of sodium hydroxide dissolved in 1 dm³ solution has a concentration of $80\,\mathrm{g\,dm^{-3}}$, $2\,\mathrm{mol\,dm^{-3}}$ or 8% **concentrate** (*v*), **concentrated** (*adj*).

dilution (*n*) (1) the process (p.157) of adding more solvent (p.86) to a solution (p.86); this lowers the concentration (↑) of the solute (p.86), e.g. the dilution of a concentrated solution of sodium hydroxide to make a dilute solution. (2) the volume in dm³ of a solution containing one mole of solute, e.g. a solution of sodium hydroxide containing $80\,\mathrm{g\,dm^{-3}}$ has a dilution of 0.5. *See diluent (p.56)*. **dilute** (*v*), **dilute** (*adj*).

concentration and dilution

1 dm³ solution
40 g sodium hydroxide
$40\,\mathrm{g\,dm^{-3}}$
1 M solution
4% solution
dilution = 1

1 dm³ solution
8 g sodium hydroxide
$8\,\mathrm{g\,dm^{-3}}$
0.2 M solution
0.8% solution
dilution = 5

analysis (*n*) a process (p.157) to find the constitution (↓) of a compound (p.8), the composition (↓) of a mixture, the concentration (p.81) of a solution, or the identity (p.225) of a substance, e.g. the analysis of the constitution of ethanoic acid; the analysis of a mixture in order to name the substances present. *See assay (p.155)*. **analyse** (*v*), **analytical** (*adj*).

volumetric analysis a way of analysis which uses solutions (p.86) of known concentrations (p.81) in titration (p.39). The reactions of the solutions may be acid-alkali, redox (p.70), or other kinds.

gravimetric analysis a way of analysis which uses reactions (p.62) forming precipitates (p.30). The precipitates are dried and weighed. Other reactions, such as the reduction (p.70) of oxides to metals may be used. In all cases, a solid is prepared (p.43) and weighed.

composition (*n*) the chemical composition of a substance gives the proportion of the elements united in the substance. The proportions can be given as a percentage composition of elements by mass, or as the proportion of atoms of each element in a covalent (p.136) molecule (p.77). The composition of a mixture gives the proportion of substances in it. **composed of** (*v*).

stoichiometric (*adj*) describes compounds which obey the law of constant proportions (p.76), or processes of analysis (↑) which measure mass or volume and assume (p.222) the law. **stoichiometry** (*n*).

structure[1] (*n*) the arrangement of connected parts of a whole with the parts depending on each other to form the structure, e.g. the structure of an atom (p.110) shows the arrangement of electrons, protons and neutrons; the structure of a crystal shows the arrangement of ions (p.123) and their dependence on each other. **structural** (*adj*).

constitution (*n*) the constitution of a còmpound (p.8) gives the number and position (p.211) in relation (p.232) to each of the other atoms forming a molecule (p.77), a radical (p.45) or an ion (p.123). The relative positions of the atoms can be shown by structural or graphic formulae (p.181). **constitute** (*v*), **constituent** (*n*).

acid

alkali + indicator

volumetric analysis with acid/alkali titration

a structure of connected and dependent parts making a whole

molecular structure a molecule consists of atoms combined by covalent bonds (p.136). A covalent bond has a direction in space, so each molecule has bonds directed in space which give the molecule a structure. The important molecular structures have the shape of a straight line, a tetrahedron ($\downarrow$), an octahedron ($\downarrow$) and a pyramid (p.84). Cations and anions (p.125) also have a molecular structure.

linear CO_2

linear (*adj*) in the shape of a straight line. A linear molecule has three atoms in a straight line, e.g. carbon dioxide, nitrogen monoxide. The central atom has no lone pairs (p.133) of electrons.

non-linear H_2O

non-linear (*adj*) in the shape of a bent line, *see diagram*. A non-linear molecule has three atoms joined by two bonds which are at an angle to each other, e.g. water has two hydrogen atoms combined with one oxygen atom and the angle between the covalent bonds is 104.5°. The central, oxygen atom has two lone pairs (p.133) of electrons.

trigonal planar CO_3^{2-}

trigonal planar (*adj*) describes a molecule with one central atom and three other atoms joined to it. All four atoms are in the same plane, e.g in the carbonate ion (CO_3^{2-}), *see p.49*, the angle between the bonds is 120° The central, carbon atom has no lone pairs (p.133) of electrons.

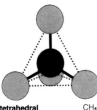

tetrahedral CH_4

tetrahedral (*adj*) in the shape of a tetrahedron, *see diagram*. A central atom is joined to four other atoms, and all the covalent bonds (p.136) have equal angles of 109.5° between each pair of bonds, e.g. methane (CH_4) has a tetrahedral structure. The central, carbon atom has no lone pairs (p.133) of electrons. **tetrahedron** (*n*).

octahedral $Fe(CN)_6^{3-}$

octahedral (*adj*) in the shape of an octahedron, *see diagram*. A central atom is joined to six other atoms, e.g. the hexacyanoferrate ions (p.53) have an octahedral structure. The central atom has no lone pairs (p.133) of electrons in an octahedral structure.

pyramidal (*adj*) in the shape of a pyramid. A pyramid in molecular structure can have a triangular base or a square base. **pyramid** (*n*).

trigonal pyramidal in the shape of a pyramid with a triangular base. A central atom is joined to three other atoms, e.g. ammonia (NH_3) has a trigonal pyramidal structure. The central, nitrogen atom has one lone pair (p.133) of electrons, which completes a tetrahedral structure.

square pyramidal in the shape of a pyramid with a square base. A central atom is joined to five other atoms. This structure is not common. The central atom has one lone pair (p.133) of electrons.

square planar in the shape of a square. A central atom is joined to four other atoms and all five atoms are in the same plane, e.g. the tetrachloroplatinate (VI) ion ($PtCl_4^{2-}$). The central, platinum atom has two lone pairs (p.133) of electrons, which complete an octahedral (p.83) structure.

trigonal bipyramidal in the shape of two triangular pyramids joined at their bases. A central atom is joined to five other atoms, e.g. phosphorus pentachloride (PCl_5) has a trigonal bipyramidal structure. The central, phosphorus atom has no lone pairs (p.133) of electrons.

NH_3 **trigonal pyramidal**

square pyramidal

$PtCl_4^{2-}$ **square planar**

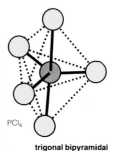

PCl_5 **trigonal bipyramidal**

SHAPE	NUMBER OF BONDS	NUMBER OF LONE PAIRS	EXAMPLE
linear	2	0	CO^2
trigonal planar	3	0	NO_3^-
tetrahedral	4	0	CH_4
trigonal pyramidal	3	1	NH_3
non-linear	2	2	H_2O
trigonal bipyramidal	5	0	PCl_5
octahedral	6	0	$Fe(CN)_6^{3-}$
square pyramidal	5	1	—
square planar	4	2	$PtCl_4^{2-}$

content (*n*) (1) the amount of a substance in a mixture, particularly the amount of a metal in a mineral, e.g. the silver content of an ore (p.154). (2) the amount a vessel contains. Usually given as the contents of the vessel.

available (*adj*) describes anything that can be obtained if it is needed, e.g. a gas is available from a generator (p.27); carbon dioxide in the air is available to plants; hydrogen in an acid is available from a chemical reaction. Anything available does not necessarily exist (p.213) uncombined.

initial (*adj*) the first part of a process, the first event in time, e.g. the first reading on a burette is the initial reading, taken before the contents (↑) are run into a flask; the mass of a substance before it undergoes (p.213) a chemical change is its initial mass; the first temperature recorded in an experiment is the initial temperature.

intermediate (*adj*) describes any part, or event, between the initial (↑) and final (↓) part or event.

final (*adj*) the last part of a process, the last event in time, e.g. the final reading on a burette after titration is finished; the final mass left after a substance has undergone (p.213) chemical change; the final temperature recorded at the end of an experiment.

qualitative (*adj*) describes an observation (p.42) in which no actual measurement is made, but quantities (p.81) are compared (p.224), e.g. warm, hot, red-hot, are qualitative observations on temperature; a rapid evolution (p.40) of hydrogen is a qualitative observation (by comparison with a slow evolution). **quality** (*n*).

quantitative (*adj*) describes an observation with a measurement, e.g. a quantitative measurement of the heat given out in a chemical reaction, measured in joules. **quantity** (*n*).

predict (*v*) to say what is going to happen in the future and to be sure it is correct, e.g. to predict the volume of a gas after a change in pressure, using Boyle's law (p.105). Predictions are made by using chemical laws. **prediction** (*n*), **predictable** (*adj*).

solute (*n*) a solid or a gas which dissolves (p.30) in a liquid.

solvent (*n*) the liquid in which a solute (↑) is dissolved (p.30).

solution (*n*) the result of dissolving a solute (↑) in a solvent (↑). If it is a homogenous (p.54) mixture of two or more substances it is called a *true solution*. Liquids other than water can be used as solvents (↑) to make a solution, but water is the usual solvent. For example, iodine can be dissolved in trichloromethane (chloroform). If a solvent other than water is used, it must be stated (p.222). A solution can also be made by dissolving one liquid in another; the liquid in greater quantity is the solvent, e.g. a solution of ethanol (ethyl alcohol) in water.

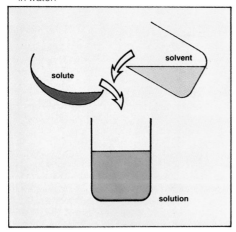

suspension (*n*) finely divided (p.13) particles of an insoluble (p.17) substance suspended (p.31) in a liquid forming an homogenous (p.54) mixture. For example, clay shaken up with water forms a suspension; lime added to water forms a solution (↑) and the excess solid forms a suspension. When a suspension is filtered, the solid substance is collected as a residue (p.31).

solubility (*n*) (1) the property of being soluble
(p.17); or a qualitative (p.85) observation on
how soluble a substance is. (2) the mass in
grams of a solid which will dissolve in 100
grams of solvent (↑) at a stated (p.222)
temperature in the presence of excess solute
(↑). (3) the volume of gas, measured in cm³,
which will saturate (↓) 100 g of solvent at a
stated temperature. Water is considered to be
the usual solvent, any other solvent must be
named. **soluble** (*adj*).

saturated

saturated
solution

excess solute
(does not dissolve)

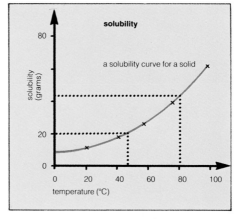

solubility

a solubility curve for a solid

saturated[1] (*adj*) (1) describes a solution (↑)
which will not dissolve any more of a solute (↑).
(2) describes air containing water vapour when
the air will not hold any more water vapour at a
particular temperature. **saturation** (*n*),
saturate (*v*).
unsaturated[1] (*adj*) describes any solution (↑), or
air, which is not saturated (↑) as more solute
(↑) can be dissolved (p.30).
supersaturated (*adj*) an unstable (p.75) state of
a solution (↑) which contains more solute (↑) at
a given temperature than it should contain. This
can happen when a warm, saturated (↑)
solution is slowly cooled. **supersaturate** (*v*),
supersaturation (*n*).

aqueous (*adj*) describes a solution (p.86) with water as the solvent (p.86), e.g. an aqueous solution of potassium hydroxide is made by dissolving potassium hydroxide in water.

non-aqueous (*adj*) describes a solution (p.86) with a solvent (p.86) other than water; the solvent is usually named, e.g. an alcoholic (ethanolic) solution of potassium hydroxide.

concentrated (*adj*) describes a solution (p.86) containing a high proportion of solute (p.86). **concentration** (*n*), **concentration** (*v*).

dilute (*adj*) describes a solution (p.86) containing a low proportion of solute (p.86). **dilution** (*n*), **dilute** (*v*).

molar concentration a term used incorrectly, but often used to mean the concentration in moles (p.80) in 1 dm³ of solution (p.86). It is better to use the word *concentration* and say mol dm^{-3}

molarity (*n*) the number of moles (p.80) of solute (p.86) in one cubic decimetre of solution (p.86), e.g. the mass of 1 mole of sodium hydroxide is 40 g. A solution of sodium hydroxide containing 4 g dm^{-3} contains 0.1 mol dm^{-3}, so the molarity is 0.1.

sodium hydroxide
4 grams
0.1 mole

dissolved in
solution

1 dm³ solution
4 g dm^{-3}
0.1 M solution
molarity = 0.1

molarity

molality

water
1 kg

solution

sodium
hydroxide
4 grams
0.1 mole

molality = 0.1

molality (*n*) a method of measuring concentration, giving the number of moles (p.80) of solute (p.86) in one kilogram of solvent (p.86).

M-value (*n*) a method of describing concentration of a solution (p.86) giving the number of moles of solute (p.86) in one cubic decimetre of solution, i.e. the value of the molarity (↑). If a solution has a molarity of 0.1 it is described as 0.1 M.

40 g is gram
molecular weight

**gram molecular
weight**

gram molecule the molecular weight of a
compound stated (p.222) in grams. It is found
by adding the atomic weights (p.114) of the
elements in the compound. This measurement
is no longer used; instead the amount of a
compound is given in moles (p.80).

equivalent weight

$$Na + H_2O \rightarrow NaOH + (\overset{\uparrow}{H})$$
23 g   40 g   1 g

23 g is equivalent weight of sodium

$$Mg + 2HCl \rightarrow MgCl_2 + (\overset{\uparrow}{H}_2)$$
24 g   2 g

12 g is equivalent weight of magnesium
(direct replacement of hydrogen by
a metal)

equivalent weight the mass of an element which
will combine (p.64) with or displace (p.68),
directly or indirectly, 1 gram of hydrogen. The
equivalent weight of an acid is the mass, in
grams, of an acid that contains 1 g of replaceable (p.68) hydrogen. The equivalent weight of
an alkali, or a base, is that mass that neutralizes
(p.67) the equivalent weight of an acid. The
equivalent weight of an oxidizing agent is the
mass of the agent which provides 8 g oxygen.
This measurement is no longer used.

normal solution

1 dm³

nitric
acid

63 g dm⁻³
1 N solution

neutralizes sodium
hydroxide
solution

1 dm³

40 g dm⁻³
1 N solution

normal solution a solution (p.86) containing the
equivalent weight (↑) of a substance dissolved
in one cubic decimetre of solution, e.g. 98 g
sulphuric acid contains 2 g replaceable
hydrogen, hence the equivalent weight of
sulphuric acid is 49. A normal solution of
sulphuric acid contains 49 g dm⁻³

normality (*n*) the fraction of the equivalent weight
(↑) of a solute (p.86) dissolved in one cubic
decimetre of solution (p.86), e.g. if 1 dm³ of
solution contains 98 g sulphuric acid, then it
contains 2 equivalent weights and its normality
is 2. This is usually written as 2N.

standard solution a solution (p.86) whose concentration is known accurately. The concentration can be given in grams per cubic decimetre, as a molarity (↑) or as a normality (↑).

solvation (*n*) a process in which molecules of a
solvent (p.86) become attached to ions (p.123),
or molecules (p.77), of solutes (p.86). **solvate** (*v*).

hydration (*n*) solvation (↑) when water is the
solvent (p.86). *See aqua-ion (p.132)*. **hydrate**
(*v*), **dehydrate** (*v*).

**hydration
(solvation)**

water molecule

copper (II) ion

adduct

water molecule

adduct (*n*) a compound (p.8) with some molecules
(p.77), usually of solvent (p.86), combined with
metal ions through weak bonds, e.g. hydrogen bonds.

hydrate (*n*) a crystal (↓) with molecules of water
combined with ions (p.123) in its crystal
structure, e.g. zinc sulphate has seven water
molecules in its crystal lattice (p.92): $ZnSO_4.$
$7H_2O$ **hydrated** (*adj*).

water of crystallization water molecules in
hydrates (↑). A particular crystalline substance
always has the same number of molecules of
water of crystallization combined with the ions
(p.123) of a molecule of that substance, e.g.
$CuSO_4.5H_2O$; $CuCl_2.2H_2O$; $Na_2SO_4.10H_2O$;
$Na_2CO_3.10H_2O$; $Na_2CO_3.H_2O$.

mother liquor

mother
liquor

crystals

mother liquor the solution left after crystals (↓)
have formed.

supernatant (*adj*) describes the liquid above a
precipitate (p.30) or a sediment (p.31). A
supernatant liquid is separated from a sediment
by decantation (p.31).

nylon
thread

**growing a
crystal**

seed
crystal

time

saturated
solution

large
crystal

crystal shapes

plane face

octahedral crystal

prismatic crystal

cubic crystal

crystallization

saturated solution

crystallization

crystal

crystallization

warm
solution

solution
cools

crystals
formed

crystal (*n*) a solid substance (p.8) with a regular
shape. It has plane (p.92) faces, which are
always at the same angle for similar sides in all
crystals of the substance. The crystal shape is
a property (p.9) of a crystalline substance.
crystalline (*adj*), **crystalloid** (*n*), **crystallize**
(*v*).

crystalloid (*n*) a substance which forms crystals
(↑) and forms a true solution (p.86) in water. A
crystalloid passes through a permeable
membrane (p.99). *See colloid (p.98).*

crystallization (*n*) the process of crystals (↑)
forming in a solution of a crystalline (↑)
substance; also the production of crystals in
an experiment (p.42).

recrystallization (*n*) to form crystals (↑) of a
substance, then to dissolve the crystals in a
solvent and to crystallize (↑) again. This makes
sure the crystals consist only of the pure
substance.

fractional crystallization a process for
separating two crystalline (↑) solids which have
solubilities of nearly the same value. The
substances undergo recrystallization (↑) many
times. After each crystallization the crystals will
be richer in one of the substances and the
mother liquor (p.90) richer in the other. At the
end of separation (p.34) pure specimens (p.43)
of both substances can be obtained.

polymorphism (*n*) the state of a solid substance (p.8) existing (p.213) in two, or more, crystalline (p.15) forms. Elements and compounds can be polymorphic. *See allotropy (p.118).* Examples are: mercury (II) oxide which has a red and a yellow crystalline form; sodium carbonate which has several hydrates, the decahydrate ($10H_2O$) forms monoclinic (p.96) crystals, the heptahydrate ($7H_2O$) forms rhombic (p.96) crystals and the transition point ($\downarrow$) between the two forms is 32°C. **polymorphic** (*adj*).

enantiotropy (*n*) polymorphism ($\uparrow$) in which two stable (p.74) crystalline forms exist, one below a transition point ($\downarrow$) and one above; the change between the two forms is reversible (p.216), e.g. sulphur has two crystalline forms, rhombic S_∞ and monoclinic S_β. S_∞ is stable below 96°C while S_β is stable above 96°C. 96°C is the transition point ($\downarrow$). **enantiotropic** (*adj*).

monotropy (*n*) polymorphism ($\uparrow$) in which there is only one stable (p.74) crystalline form and other forms are unstable, e.g. red phosphorus is the stable form of phosphorus and white phosphorus is unstable. There is no transition point ($\downarrow$) between the two forms. **monotropic** (*adj*).

transition point (*n*) the temperature at which one crystalline (p.15) form of a substance changes to another crystalline form in a reversible (p.216) change. At the transition temperature, both crystalline forms can exist (p.213) together.

lattice (*n*) a regular arrangement of points in space with a pattern ($\downarrow$) that can be recognized.

crystal lattice (*n*) a lattice ($\uparrow$) with atoms, molecules, or ions at the points of the lattice. A lattice has a pattern in three dimensions and reaches to the faces of the crystal (p.91).

isomorphism (*n*) the state of crystals of two different substances having the same crystal lattice ($\uparrow$), hence the same crystal shape. **isomorphic** (*adj*).

plane (*n*) a flat surface. Two planes meet in a straight line. Three planes can meet in the same straight line, or at a point. **planar** (*adj*).

axis (*n*) (*n.pl. axes*) (1) a line drawn through an object which divides it into two equal halves. (2) a line about which an object turns. **axial** (*adj*).

stable below 96°C
rhombic crystal
S_∞
S_β
96°C
transition point
stable above 96°C
monoclinic crystal

enantiotropy of sulphur

lattice

two planes meet in a line

symmetry

axis of symmetry
point of symmetry
plane of symmetry

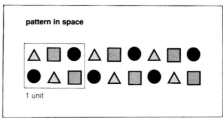

pattern in space

1 unit

orientation

90°

orientation of 90° to other line

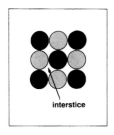

interstice

pattern (*n*) a regular arrangement of points, objects, or shapes which recur (p.217) in space; each arrangement is called a unit of the pattern. Processes or events which occur regularly in time form a dynamic pattern.

symmetry (*n*) the state of having a regular shape such that a line or plane can be drawn which divides the shape into two equal halves, e.g. a circle has symmetry as any line through the centre divides it into two equal halves. **symmetrical** (*adj*), **asymmetrical** (*adj*).

crystal symmetry symmetry (↑) of the lattice (↑) or shape of a crystal (p.91). Crystals can have planes, lines or centres of symmetry.

crystal face (*n*) a plane surface of a crystal.

plane of symmetry a plane which divides the lattice (↑) or the shape of a crystal into two equal halves.

orientation (*n*) the direction in which an object points in relation to its surroundings (p.103), e.g. the orientation of one line in relation to another line which it meets at a point.

perfect crystal (*n*) a crystal (p.91) which has no breaks in the lattice (↑), i.e. no atoms, ions or molecules missing from the lattice.

irregular (*adj*) describes any arrangement, in space or in time, which is not regular. The opposite of regular.

interstice (*n*) a small space between the ions, atoms or molecules in a crystal lattice (↑), e.g. in steel, there are atoms of carbon in the interstices of the iron atoms forming the lattice of the metal crystals. *See metal crystals (p.95).* **interstitial** (*adj*).

cleavage plane a plane in a crystal (p.91) along which a blow will separate the crystal into two parts with a clean cut. The crystal is *cleaved* by the blow. Along any other plane the crystal will be shattered (↓). **cleave** (*v*), **cleavage** (*n*).

ions

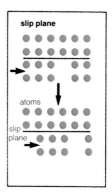

cleavage plane

shatter (*v*) to break a hard, or brittle (p.14) object into very many small pieces by a hard blow, e.g. a cup dropped on the ground is shattered.

slip plane a plane in a crystal (p.91), particularly a metal crystal (↓), along which one part of the crystal will move relative to the other part, when a force acts on the crystal. One part of the crystal lattice *slips* over the other part. Slip planes in metal crystals cause metals to be *ductile* (p.14).

giant structure a crystal lattice (p.92) with no separate molecules (p.77). The ions (p.123) or atoms (p.110) forming the lattice are all joined, one to another, by bonds (p.133), so that the crystal appears to be one very large molecule. Many giant structures are held together by ionic bonds (p.134).

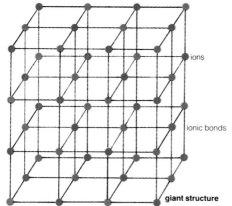

ions

ionic bonds

giant structure

molecular crystal a crystal consisting (p.55) of separate molecules held together by weak van der Waals' (p.137) forces. The crystal is very soft and has a low melting point (p.12). Many organic compounds form molecular crystals.

giant molecular structure   copper (II) chloride

atoms
copper (II)

chlorine

bonds
covalent
co-ordinate

close packing

close packing of atoms

hexagonal close packing

● top layer
◐ middle layer
○ bottom layer

cubic close packing

● top layer
◐ middle layer
○ bottom layer

giant molecular crystal a crystal with a giant structure (↑) composed of atoms joined by covalent bonds (p.136). A giant molecular crystal is very hard and has a very high melting point (p.12). Aluminium oxide and anhydrous copper (II) chloride form giant molecular crystals.

metal crystal a giant structure (↑) formed by metals. The crystal lattice consists (p.55) of positive ions of metals. A cloud of electrons can move throughout the lattice and holds the positive ions together. The cloud of freely moving electrons accounts for the ability of metals to conduct electric current (p.122) and heat.

close packing an arrangement of atoms put as close to each other as possible. This forms a close packed lattice of one layer (p.18). Two arrangements of close packing are found in metal crystals (↑), hexagonal (↓) and cubic (↓).

hexagonal close packing a second layer of atoms lies in the hollows formed by groups of three atoms in the first layer of close packing (↑). A third layer of atoms similarly lies in the hollows formed by groups of three atoms in the second layer; each atom in the third layer is directly below an atom in the first layer.

cubic close packing the atoms in the first two layers are arranged in the same way as the atoms in hexagonal close packing (↑). The third layer of atoms lies in the hollows formed by groups of three atoms, but these atoms are not below the atoms in the first layer.

crystal systems crystals are grouped into
systems according to their symmetry (p.93).

cubic system a system with three equal axes of
symmetry (a = b = c). The angle between any
two axes is 90°.

tetragonal system a system with two equal axes
of symmetry and a third axis which is longer or
shorter (a = b ≠ c). The angle between any two
axes is 90°

orthorhombic system a system with three
unequal axes of symmetry (a ≠ b ≠ c). The
angle between any two axes is 90°.

a = b = c   **cubic system**

tetragonal system

a = b ≠ c

monoclinic
system

a ≠ b ≠ c

aÔb ≠ 90°

monoclinic system a system which has three
unequal axes of symmetry. The angle between
two pairs of the axes is 90°, the angle between
the third pair of axes is not 90° but either greater
or smaller.

triclinic system a system which has three
unequal axes of symmetry. The angle between
any two axes is not 90°.

hexagonal system a system which has three
equal axes of symmetry at an angle of 120° to
each other with a fourth axis of symmetry at an
angle of 90°, and of a different length.

orthorhombic system

a ≠ b ≠ c

triclinic system

no angle of 90°

a ≠ b ≠ c

hexagonal system

$a_1 = a_2 = a_3 \neq c$

$a_1Ôa_2 = a_2Ôa_3 = a_3Ôa_1 = 120°$

$cÔa_1 = cÔa_2 = cÔa_3 = 90°$

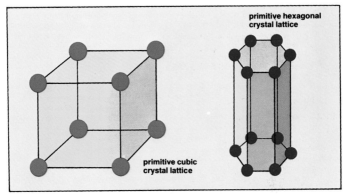

primitive hexagonal crystal lattice

primitive cubic crystal lattice

primitive structure this is the simplest crystal structure with an atom (p.110) or ion (p.123) at each
corner of the figure for each of the crystal structures. *See diagram*, which shows a primitive
cubic lattice and a primitive hexagonal lattice.

body-centred lattice this structure is a primitive
structure (↑) with an additional atom or ion at
the centre of the figure.

face-centred lattice this structure is a primitive
structure (↑) with additional atoms or ions at
the centre of each face of the figure.

body-centred lattice

face-centred lattice

colloids
effect of a light beam on
true solutions and colloids

colloid (*n*) a constituent (p.54) of a disperse (↓) system (p.212) in which that constituent is dispersed throughout another constituent, e.g. a constituent (clay) is dispersed in water (the other constituent). The colloidal constituent is present (p.217) in particles which are between 1 nm and 100 nm in size. If the colloid is a solid and it is dispersed in a liquid, the size of the particles (p.13) is greater than those in a true solution (p.86), but smaller than those in a suspension (p.86). Solid colloidal particles dispersed in a liquid pass through a filter (p.30) but not through a permeable membrane (↓). Examples of colloids include: milk (a dispersion of colloidal particles of fat in water); foam (p.100); aerosol (p.100); sol (p.100); froth (p.100); smoke (p.100). **colloidal** (*adj*).

disperse (*v*) to spread particles (p.13), or similar small objects, over a large area or volume. The action points to the agent causing it, e.g. a liquid disperses a solid colloid (↑) over all its volume. **dispersion** (*n*), **disperse** (*adj*), **dispersed** (*adj*).

dispersion medium the liquid, or gas, that disperses (↑) a colloid (↑) over all its volume. It has a continuous phase whereas the disperse phase is discontinuous.

disperse phase the colloidal (↑) substance that is dispersed (↑) over all the dispersion medium (↑).

membrane (*n*) a thin piece of material such as parchment paper, cellophane, a bladder. A liquid can pass slowly through a membrane if the membrane is permeable (↓). A crystalloid (p.91) solid can also pass through a permeable membrane, but a colloidal (↑) solid and a suspension cannot pass through it.

permeable (*adj*) describes any object which allows fluids (p.11) to pass through by diffusion (p.35). Permeable membranes (↑) allow molecules (p.77) of fluids to diffuse. Other substances which are finely divided, e.g. sand, also allow particles (p.13) to diffuse. **permeability** (*n*).

coagulate (*v*) to cause small particles to join together to form bigger particles or a mass of particles, e.g. a colloidal sol (p.100) can be coagulated to form a suspension (p.86); an iron (III) oxide sol in water is coagulated to a suspension by a solution of aluminium chloride. **coagulation** (*n*), **coagulated** (*adj*).

sol (*n*) a sol is formed by a colloidal (p.98) solid dispersed (p.99) in a liquid. The colloidal solid passes through a filter paper (p.30) but does not pass through a permeable membrane (p.99). Crystalloids (p.91) are separated from colloids by dialysis (p.34). Sols are described as lyophilic or lyophobic (↓).

sol

colloidal solid dispersed in liquid

hydrosol (*n*) a sol (↑) in which the dispersion medium (p.99) is water.

emulsion (*n*) an emulsion is formed from two liquids, one of which is dispersed (p.99) in the other liquid in the colloidal state. If the emulsion is not stabilized (↓) the liquids separate out, e.g. coconut oil and water form an emulsion when shaken together; the two liquids separate on standing. Soap acts as an emulsifying agent (↓) to stabilize the emulsion. Milk is an emulsion with butter fat dispersed in a dilute sugar solution. **emulsify** (*v*), **emulsified** (*adj*).

emulsion

oil

emulsion

water

foam (*n*) a foam is formed from a gas dispersed (p.99) in a liquid. It consists of small bubbles of gas with a thin film (p.18) of liquid round the bubbles. A foam is usually stable. Foams are formed when a gas and a liquid are forced through a jet under pressure. **foam** (*v*), **foaming** (*adj*).

foam

foam

gas dispersed in liquid

liquid

froth (*n*) a foam (↑) which has large bubbles and is less stable. Mixtures of ethanol and water form a froth, with carbon dioxide as the gas. **froth** (*v*).

aerosol (*n*) in an aerosol, a liquid is dispersed in a gas, usually air. It is formed by atomization (↓). A mist is an example of an aerosol. Many insecticides are used as aerosols.

liquid dispersed in air

aerosol

smoke (*n*) smoke is formed by solids dispersed in a gas, usually air, e.g. small particles (p.13) of carbon dispersed in air in wood smoke.

gel (*n*) an intermediate (p.85) stage (p.159) between a sol (↑) and a suspension. The colloidal particles (p.13) form long thin threads round the liquid dispersion medium (p.99). The result appears to be solid but it is easily deformed. A jelly is a gel. **gel** (*v*), **gelation** (*n*).

smoke

solid dispersed by gas

smoke

lyophilic (*adj*) describes a sol (↑) in which the disperse phase (p.99), a solid, has an attraction (p.124) for the dispersion medium (p.99), a liquid. The colloid (p.98) goes readily into solution; after coagulation (p.99), it is readily dispersed (p.99) again by adding more liquid, e.g. a starch sol is lyophilic.

lyophobic (*adj*) describes a sol (↑) in which the disperse phase (p.99), a solid, has a repulsion (p.124) for the dispersion medium (p.99), a liquid. The colloid (p.98) readily comes out of solution; after coagulation (p.99) it does not become dispersed (p.99) again, e.g. a gold sol.

hydrophilic (*adj*) describes a lyophilic (↑) sol with water as the dispersion medium (p.99).

hydrophobic (*adj*) describes a lyophobic (↑) sol with water as the dispersion medium (p.99).

stabilize (*v*) to make a sol, emulsion, or gel, stable, by preventing them from changing to a suspension, or from coagulating or from separating. **stable** (*adj*), **stabilizer** (*n*), **stabilization** (*n*).

emulsifying agent an agent (p.63) which stabilizes (↑) an emulsion, e.g. soap stabilizes a coconut oil and water emulsion.

atomize (*v*) to blow a liquid through a fine (p.13) jet (p.29). This forms an aerosol (↑) with the liquid dispersed (p.99) in air. **atomizer** (*n*).

cataphoresis (*n*) the movement of colloidal (p.98) particles (p.13) towards electrodes in a lyophobic (↑) sol, or in a smoke (↑). Lyophilic (↑) sols may undergo (p.213) cataphoresis if the colloid has an electric charge. The colloidal particles may move either to the anode (p.123) or the cathode (p.123) depending on their charge, e.g. clay and metallic particles are negatively charged; iron (III) chloride and aluminium hydroxide are positively charged in aqueous (p.88) sols. Platinum electrodes are usually used in cataphoresis.

electrophoresis (*n*) another name for cataphoresis (↑).

thixotropy (*n*) a property of some colloidal liquids by which the liquid has a decreased viscosity at an increased rate of flow. **thixotropic** (*adj*).

atomizer

aerosol

cataphoresis

platinum
electrodes
cathode   anode

water

moves
upwards

gold sol

pressure (*n*) the force per unit area acting on a surface. It is the force in newtons acting on an area measured in square metres. A force of 200 newtons acting on an area of 5 sq metres gives a pressure of $200\,N \div 5\,m^2 = 40\,N\,m^{-2}$. The unit of pressure is the pascal (Pa); a pressure of $1\,N\,m^{-2} = 1\,Pa$. Pressure in gases is also measured by the height of a column of mercury supported by the pressure; the height is given in millimetres (mm) of mercury. 1 mm of mercury = $133.322\,N\,m^{-2}$. **press** (*v*).

pressure = 60 Nm⁻²

atmospheric pressure the pressure (↑) exerted (p.106) by the air in the atmosphere on the Earth's surface. It is measured by a barometer.

pressure = 60 Nm⁻²

standard atmospheric pressure an atmospheric pressure (↑) of 101 325 N m⁻² or 760 mm of mercury. The everyday variations of atmospheric pressure are above and below this value.

s.t.p. (*abbr*) an abbreviation for standard temperature (↓) and pressure (↑), i.e. 0°C or 273 K and 101 325 Nm⁻²

temperature (*n*) a physical property which determines the direction of the flow of heat between materials (p.8) in thermal (p.65) contact (p.217). Heat flows from a higher to a lower temperature.

Celsius scale a temperature scale (p.26) with 0° as the temperature of melting ice and 100° as the temperature of steam at standard atmospheric pressure (↑). One Celsius degree (1°C) is 1/100 of the temperature interval (p.220) between the ice point and the steam point. The symbol for Celsius temperature is θ.

absolute scale a temperature scale (p.26) with 0 as the absolute zero of temperature and a temperature interval of 1 kelvin. On this scale the temperature of melting ice is 273 K and of steam is 373 K. Absolute temperatures are changed to Celsius temperatures by: 273 + Celsius temperature = absolute temperature. The symbol for absolute temperature is T.

kelvin (*n*) the S.I. unit of temperature, measured from the triple point of water at which ice, water and water vapour all exist (p.213) together. One kelvin is 1/273.16 of the value of the triple point. A temperature interval of 1 K = 1°C (symbol = K).

surroundings (*n.pl.*) all the objects and materials near to and around an object, or the objects and materials for an experiment, are the surroundings of that object or experiment. The surroundings may, or may not, have an effect on the object or experiment. **surround** (*v*).

factor (*n*) a possible cause of an effect or change. Some of the surroundings (↑) are factors, i.e. those that have an effect on an object or experiment. Factors for chemical reactions are temperature (↑), pressure (↑), concentration (p.81), catalysts (p.72), surface nature (p.19).

conditions (*n.pl.*) the magnitude of factors (↑) of a particular object or experiment. To contrast *factors* and *conditions*: (1) temperature is a *factor* for the rate of reaction (p.149) between substances; a temperature of 60°C is an actual *condition* in an experiment. (2) water is a *factor* in the rusting (p.61) of iron; the presence of water is a necessary *condition*. Conditions can be *necessary*, i.e. without them, there is no reaction or effect, or conditions may be *adverse*, under which the reaction or effect does not take place or takes place very slowly, or conditions may be *suitable*, under which the reaction or effect is normal (p.229).

mercury barometer

using a mercury barometer to measure vapour pressure of a liquid

change of vapour pressure with temperature

vapour pressure the pressure (↑) exerted (p.106) by a saturated (p.87) vapour (p.11). Vapour pressure increases with temperature. A liquid boils when its vapour pressure is equal to the atmospheric pressure (↑).

ambient (*adj*) describes conditions of the surroundings (↑), e.g. the ambient temperature is the temperature of the surrounding air.

condensation[2] (*n*) a change of state (p.9) from vapour, or gas, to liquid caused by a decrease in temperature. **condense** (*v*).

liquefaction (*n*) a change of state (p.9) from vapour to liquid caused by an increase in pressure. **liquefy** (*v*).

permanent gas a gas which is not easily liquefied because it has to be first cooled to a very low temperature, e.g. hydrogen, nitrogen and oxygen are permanent gases.

noble gas one of the following gases: helium, neon, argon, krypton, xenon, and radon. They do not take part in chemical reactions and are usually considered to be inert (p.19). They are permanent gases (↑) and also monatomic (↓).

critical temperature a particular temperature for each gas, above which the gas cannot be liquefied (↑) by pressure (p.102) alone. The critical temperature for oxygen is −119°C (154 K), so oxygen has to be cooled below −119°C before it can be liquefied by pressure; for this reason, oxygen is a permanent gas (↑). Critical temperature has the symbol T_c.

critical pressure (*n*) the pressure sufficient to liquefy a gas at its critical temperature (↑).

monatomic (*adj*) describes a gas with molecules (p.77) consisting of one atom (p.110) only, e.g. neon and helium are monatomic gases.

monatomic   neon (Ne)

diatomic (*adj*) describes a gas with molecules (p.77) consisting of two atoms (p.110).

polyatomic (*adj*) describes a gas with molecules (p.77) consisting of several atoms (p.110). Such gases are usually compounds, e.g. carbon dioxide.

diatomic   hydrogen (H_2)

atomicity (*n*) the number of atoms in a molecule of an element or a compound, when the element or compound consists of separate molecules. In most cases, the substances will be gaseous, e.g. helium has an atomicity of 1; hydrogen of 2; sulphur dioxide of 3.

gram-molecular volume the volume of one gram-molecule (the molecular weight in grams, calculated from the sum of the atomic weights in grams) of a substance. Molar volume is now used instead of gram-molecular volume.

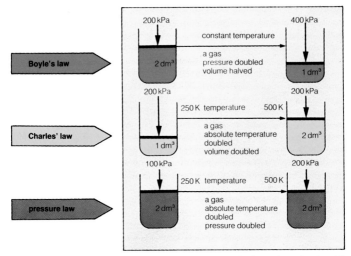

Boyle's law the volume of a fixed (p.79) mass of
gas is inversely proportional to the pressure
(p.102) at a constant (p.106) temperature; the
relation can also be written as pressure times
volume equals a constant. In symbols $pV = k$,
where k is a constant. The law is obeyed
(p.107) at low pressures, but real (p.107)
gases do not obey the law at high pressures.

Charles' law the volume of a given mass of gas
increases by 1/273 of its volume at 0°C for
each degree Celsius rise in temperature, if the
pressure (p.102) remains constant. In symbols
$$V_\theta = V_0 (1 + \boldsymbol{\theta}/273)$$
The volume of the gas is proportional to the
absolute temperature (p.102), so $V \propto T$ and
hence $V_1/V_2 = T_1/T_2$.

pressure law the pressure (p.102) of a given
mass of gas increases by 1/273 of its pressure
at 0°C for each degree Celsius rise in
temperature, if the volume remains constant.
In symbols $p_\theta = p_0 (1 + \boldsymbol{\theta}/273)$.
The pressure of the gas is proportional to the
absolute temperature (p.102) so $p \propto T$ and
hence $p_1/p_2 = T_1/T_2$.

gas equation the three laws, Boyle's, Charles'
and pressure, can be combined into one
equation which is $pV = kT$, where k is a
constant ($\downarrow$) depending on the mass of gas.
The equation is usually written as $p_1V_1/p_2V_2 = T_1/T_2$. If the amount of gas is given as a mole
fraction (p.80) with symbol n, then the equation
becomes $pV = nRT$, where R is the same
constant for all gases.

constant[1] (n) (1) the value of a physical quantity
(p.81) which cannot be changed, e.g. the speed
of light is a constant; the Avogadro constant
(p.80). (2) the unchanging value of a physical
quantity under experimental conditions (p.103),
e.g. the boiling point of a liquid at s.t.p. (p.102).

constant[2] (*adj*) describes an unchanging value
of a physical quantity which is controlled
(p.221) by an observer.

graph of volume against
pressure (proof of Boyle's law)

curve

volume

pressure

curve (n) the line drawn in a graph showing the
relation (p.232) between two quantities (p.81).

derive (v) to obtain a value or a statement
(p.222) by several steps of experimental work
or deduction (p.222). The work must start from
something initially (p.85) known. For example,
Boyle's law is derived from experiments on
gases, measuring volumes and pressures,
plotting (p.39) graphs and then deducing the
relation between pressure and volume. The
statement of the law is then derived.
derivation (n), **derivative** (n), **derived** (*adj*).

exert (v) to bring into effect, e.g. a gas exerts a
pressure on the walls of a vessel (p.25) so the
pressure has an effect on the walls.

ideal (*adj*) (1) describes a gas which obeys (↓) Boyle's law (p.105) at all pressures. Such gases do not exist (p.213). Real (↓) gases are ideal gases at low pressures. (2) describes an object, arrangement or theory (p.76) that is perfect; such an object, arrangement or theory is usually part of a hypothesis (p.108).

non-ideal (*adj*) (1) describes a gas which does not obey Boyle's law at all pressures. (2) describes any object or arrangement which is not ideal (↑).

real (*adj*) (1) describes any object that exists (p.213), in particular any gas that exists, e.g. gases such as hydrogen, oxygen and nitrogen are all real gases. (2) describes any object that is what it is said to be, e.g. a bag made of real leather not *imitation* leather.

conform (*v*) to follow a pattern (p.93) of behaviour, e.g. gases conform with Boyle's law, but the degree (p.227) of conformity becomes less at high pressures. A material or substance (p.8) can conform with a law (p.109), or any other statement (p.222) of its properties, but the agreement between experimental (p.42) results (p.39) and the prediction (p.85) is not necessarily exact. **conformity** (*n*).

obey (*v*) to be in exact agreement with a law, so that experimental (p.42) results (p.39) are in exact agreement with predictions (p.85), e.g. an ideal (↑) gas obeys Boyle's law; chemical compounds obey the law of constant proportions (p.76). To contrast *conform* and *obey*: if experimental results are approximately (p.79) in agreement with predictions, then a substance *conforms* to the law, if the results are in exact agreement, a substance *obeys* the law.

Graham's law the rate of diffusion (p.35), or effusion (p.35), of a gas at a constant temperature is inversely (p.233) proportional (p.76) to the square root of its density (p.12). The greater the relative molecular mass (p.114) of a gas, the slower is its rate of diffusion, or effusion.

porous plug

gas

Graham's law

A

B

20 seconds later ↓

gas effuses

A

B

rate of effusion = $\dfrac{\text{volume AB}}{20 \text{ seconds}}$

Avogadro's hypothesis equal volumes of all
gases at the same temperature and pressure
contain the same number of molecules. *See
Avogadro constant (p.80)*. The molar volume
(p.80) of a gas at s.t.p. (p.102) contains the
Avogadro number of molecules.

Avogadro's principle another name of
Avogadro's hypothesis.

hypothesis (*n*) a statement which cannot be
proved experimentally (p.42). From it, however,
laws can be deduced (p.222) and proved.

s.t.p.
22.4 dm³
6.02×10^{23} molecules

s.t.p.
22.4 dm³
6.02×10^{23} molecules

Avogadro's hypothesis

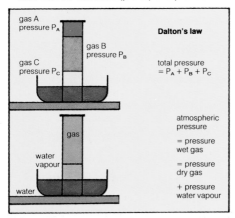

gas A
pressure P_A

gas B
pressure P_B

gas C
pressure P_C

Dalton's law

total pressure
$= P_A + P_B + P_C$

gas

water
vapour

water

atmospheric
pressure

= pressure
wet gas

= pressure
dry gas

+ pressure
water vapour

Dalton's law in a mixture of gases which do not
react (p.62), the total pressure exerted (p.106)
by the mixture of gases is equal to the sum of
the partial pressures (↓) of each gas. Also
called **Dalton's law of partial pressures**.

partial pressure the partial pressure of a gas in
a mixture of gases is equal to the pressure
which each gas would exert (p.106) if it alone
filled the whole volume of the mixture.

kinetic theory a theory (p.76) which explains the
behaviour of liquids and gases by considering
them to consist (p.55) of molecules (p.77) in
random (p.223) motion; the molecules hit each
other and the walls of the containing vessel
(p.25). The molecules are considered to be
completely elastic (p.14).

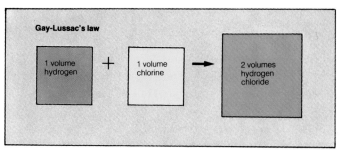

Gay-Lussac's law

| 1 volume hydrogen | + | 1 volume chlorine | → | 2 volumes hydrogen chloride |

Gay-Lussac's law the volumes of gases which react, and the volumes of the products if gaseous, are in the ratio (p.79) of small whole numbers if pressure and temperature are kept constant. For example, one volume of hydrogen combines (p.64) with one volume of chlorine to form two volumes of hydrogen chloride. One and two are simple whole numbers. *See* Charles' Law (p.105).

law (*n*) a statement about the properties of materials and substances (p.8) that is accepted as true. From it, predictions (p.85) about the behaviour of substances can be made. A law is universal (p.212) or limited (↓), e.g. the law of constant proportions (p.76) is a universal law as no substances are excepted from it; Boyle's law is a limited law as it depends on the condition of temperature being constant.

limited[1] (*adj*) describes a law (↑), or theory (p.76), which is true only under particular conditions. **limitations** (*n*), **limit** (*v*).

gas laws statements about the behaviour of gases under particular conditions. The important laws are those of Boyle (p.105), Charles (p.105) and Dalton (↑).

Brownian motion the random (p.223) motion of molecules in liquids (p.10) and gases. The motion is seen when pollen grains are suspended (p.31) in water, or when smoke particles are seen in air. In both cases, the motion of molecules causes the motion of the pollen grains or smoke particles.

atom (*n*) the smallest particle (↓) of an element
(p.8) that has the properties of that element.
Atoms combine chemically to form molecules
(p.77). An atom consists of a nucleus (↓) with
electrons (↓) around it. **atomic** (*adj*).

subatomic (*adj*) describes particles (↓) smaller
than an atom (↑).

electron (*n*) a subatomic (↑) particle with a
negative electric charge (p.138). It is the
smallest subatomic particle with a mass of
9.109×10^{-31} kg.

proton (*n*) a subatomic (↑) particle with a
positive charge (p.138). The charge is equal
and opposite to that of an electron (↑). The
mass of a proton is 1840 times that of an
electron.

neutron (*n*) a subatomic (↑) particle with a mass
almost equal to the mass of a proton (↑). It has
no electric charge (p.138).

nucleus (*n*) the nucleus of an atom (↑) consists
(p.55) of protons (↑) and neutrons (↑), except
for hydrogen, which has a nucleus consisting
of one proton. The nucleus has a positive electric charge, the size of which depends upon
the number of protons it contains. The nucleus
provides the mass of an atom. **nuclear** (*adj*).

nuclear (*adj*) describes anything to do with the
nucleus (↑) of an atom.

particle[2] (*n*) a piece of substance so small that it
is considered to have mass but not size or
volume, that is, it is like a point. Atoms (↑) and
molecules (p.77) are particles in this sense.

positron (*n*) a subatomic (↑) particle with a mass
the same as the mass of an electron (↑), but
with a positive charge equal and opposite to
the charge on an electron.

meson (*n*) subatomic (↑) particles with a mass
about half that of a proton (↑) and with zero
spin; they appear to bind nuclear particles
together.

orbit (*n*) in the first models (p.223) of atoms, the
electrons (↑) were considered to be in orbits
round the nucleus (↑). An orbit is a circular
path round a nucleus. This idea is no longer
considered to be true.

electrons around
nucleus

nucleus

structure of an atom
not to scale

orbit

orbit

shell (*n*) a spherical (↓) space around a nucleus (↑); it contains extranuclear (p.113) electrons with energy levels (p.152) of that shell. An atom possesses different shells of increasing radii (↓) around its nucleus (↑). The radius of a shell is 100 000 times greater than the radius of the nucleus.

sphere (*n*) a solid in the shape of a ball. **spherical** (*adj*).

radius (*n*) (*radii n.pl.*) (1) the line joining the centre of a circle to the line of the circle; the line drawn from the centre of a sphere to its surface. (2) the length of the line described above, e.g. a circle with a radius of 5 cm.

orbital (*n*) a space in which there can be one or two electrons (↑) but not more. The space is where there is a probability (p.223) of finding one or both electrons. The probability varies (p.218) over the space, and can be shown in various ways by a diagram (p.29). The position and motion of an electron in an orbital cannot be described, only the probability can be described. One way of showing probability is by shading, *see diagram*. Another way is by drawing the limits (p.211) of the orbital. Two electrons in an orbital do not interfere (p.216) with each other; two electrons form a stable (p.74) orbital, while one electron in an orbital is active in forming chemical bonds (p.133). The orbitals commonly taking part in bonds are s-, p- and d-orbitals (p.112).

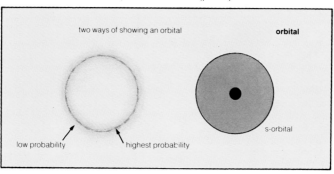

two ways of showing an orbital   **orbital**

low probability   highest probability   s-orbital

s-orbital (*n*) an orbital (p.111) in an atomic shell (p.111) containing one or two electrons (p.110). There is only one s-orbital in a shell. An s-orbital is spherically (p.111) symmetrical (p.93) about a nucleus. It has no directional properties.

p-orbital (*n*) an orbital (p.111) containing one or two electrons (p.110). There are three p-orbitals in an atomic shell (p.111). The three orbitals are at right angles to each other along three axes (p.92) and have directional properties.

d-orbital (*n*) an orbital (p.111) containing one or two electrons (p.110). There are five d-orbitals in an atomic shell (p.111). Four of the orbitals have the same shape, and one has a different shape. All five orbitals have directional properties.

s-electron (*n*) an electron (p.110) in an s-orbital (↑). s-electrons have the least energy in a shell (p.111). Two s-electrons fill the s-orbital before an electron goes into a p-orbital.

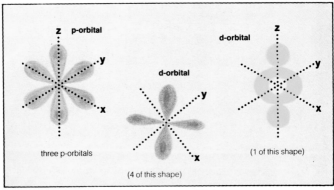

p-electron (*n*) an electron (p.110) in a p-orbital (↑). One p-electron goes into each of the three p-orbitals before the fourth electron goes into one p-orbital to fill it. A total of six p-electrons fill the three p-orbitals before an electron goes into the next higher energy level.

d-electron (*n*) an electron (p.110) in a d-orbital (↑). One d-electron goes into each of the five d-orbitals before the sixth electron goes into one d-orbital to fill it.

atomic structure

extranuclear (*adj*) describes electrons (p.110) outside the nucleus (p.110) of an atom.

atomic structure the structure of the nucleus (p.110) and of the extranuclear (↑) electrons. The extranuclear electrons are arranged in atomic shells (p.111).

K-shell (*n*) the innermost atomic shell (p.111). It contains only one s-orbital. This arrangement completes the shell.

L-shell (*n*) the next atomic shell to the K-shell (↑). It contains one s-orbital and three p-orbitals. This arrangement completes the shell.

M-shell (*n*) the next atomic shell to the L-shell (↑). It contains one s-orbital, three p-orbitals and five d-orbitals. This completes the shell. There are further shells beyond the M-shell.

atomic number a number equal to the number of protons (p.110) in the nucleus (p.110) of an atom of an element. It has the symbol Z.

mass number a number equal to the sum of the number of protons (p.110) and number of neutrons (p.110) in the nucleus (p.110) of an atom of an element. It is the mass number of an isotope (p.114). The symbol for mass number is A.

relative atomic mass the ratio (p.79) of the mass of one atom of an element to one-twelfth of the mass of one atom of carbon-12. Most elements consist of isotopes (p.114) of its atoms, hence the relative atomic mass measures the average mass per atom of the normal isotopic composition of an element.

mass number

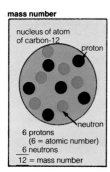

nucleus of atom
of carbon-12

proton

neutron

6 protons
(6 = atomic number)
6 neutrons
12 = mass number

isotope (*n*) atoms of the same element (p.116) which have different mass numbers (p.113) are isotopes of the element. As all atoms of an element have the same number of protons (p.110), isotopes differ in having different numbers of neutrons (p.110) in the nucleus (p.110) of the atoms, e.g. there are two isotopes of carbon, one with a mass number of 12 and one with a mass number of 13. Both isotopes have six protons in their nuclei, so one isotope has six neutrons and the other seven neutrons. Isotopes are shown by writing the mass number after the element, e.g. carbon-12 is the isotope of carbon with a mass number of 12. **isotopic** (*adj*).

isotopic ratio the ratio (p.79) of the different isotopes (↑) in a specimen (p.43) of an element. For elements obtained from natural sources (p.138) the isotopic ratio is always the same, e.g. the isotopic ratio for carbon is 98.9% C-12 and 1.1% C-13. The relative atomic mass of carbon is thus slightly greater than 12.

relative isotopic mass (*n*) (of an isotope) the ratio of the mass of one atom of the isotope to 1/12 of the mass of one atom of carbon-12.

isotopic weight (*n*) another name for relative isotopic mass (↑), no longer used.

atomic mass unit a mass equal to one-twelfth of the mass of an atom of the isotope (↑) carbon-12. Its value is 1.66043×10^{-27} kg. The symbol for atomic mass unit is *amu*.

atomic weight (*n*) the ratio of the mass of one atom of an element to one-sixteenth of the mass of an atom of oxygen. Relative atomic mass is now used instead of atomic weight, and the method of measuring is different with a standard of carbon-12 instead of oxygen.

relative molecular mass (*n*) the ratio (p.79) of the mass of one molecule of a substance (p.8) to the mass of one atom of carbon-12. The relative molecular mass of a compound is calculated by adding the relative atomic masses of all the atoms in a molecule of the compound. Relative molecular mass is used with covalent compounds, but not with ionic compounds. *See relative formula mass (p.78).*

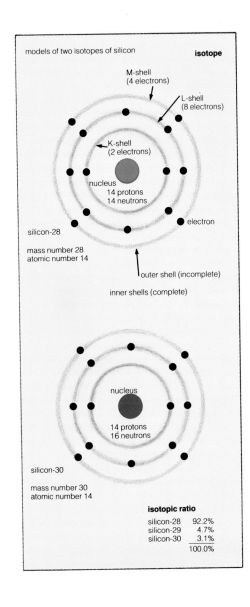

models of two isotopes of silicon **isotope**

M-shell
(4 electrons)

L-shell
(8 electrons)

K-shell
(2 electrons)

nucleus
14 protons
14 neutrons

electron

silicon-28

mass number 28
atomic number 14

outer shell (incomplete)

inner shells (complete)

nucleus

14 protons
16 neutrons

silicon-30

mass number 30
atomic number 14

isotopic ratio

silicon-28	92.2%
silicon-29	4.7%
silicon-30	3.1%
	100.0%

element[2] (*n*) a substance (p.8) with all its atoms (p.110) having the same positive charge on the nucleus (p.110), i.e. the nucleus of all atoms has the same number of protons (p.110) and thus the same atomic number, and this determines the chemical nature of the element. **elementary** (*adj*).

metal (*n*) an element (↑) which forms positive ions (p.123) in chemical reactions. The general physical properties of metals are: (a) they conduct (p.122) electric current and heat; (b) they are lustrous (p.16), ductile (p.14) and malleable (p.14). The general chemical properties of metals are: (a) they form basic (p.46) oxides; (b) they form compounds with non-metals which are salts. Metals possess these properties in varying (p.218) degrees (p.227). All metals, except mercury, are solids. **metallic** (*adj*).

non-metal (*n*) an element (↑) which is not a metal (↑). Non-metals are solids or gases, except bromine which is a liquid. Their physical properties depend upon their structure (p.82). Solid non-metals are neither ductile (p.14) nor malleable (p.14), but are usually brittle (p.14). They do not conduct electric current, except for graphite (p.118), nor heat. They generally form negative ions (p.123) in chemical reactions, except for hydrogen which forms positive ions. Their oxides are generally acidic, but some are neutral, e.g. carbon monoxide.

elements	metals		non-metals	
solids	iron	gold sodium copper	sulphur	phosphorus
liquids	mercury		bromine	
gases	none		oxygen	hydrogen nitrogen

metalloid (*n*) an element with some of the properties of metals and some of the properties of non-metals, e.g. antimony and arsenic are metalloids.

alkali metal a metal (↑) which forms an ion (p.123) with an oxidation state (p.135) of +1 only. The oxide is very soluble in water forming a hydroxide (p.48). Alkali metals are chemically very reactive (p.62); they are the elements of group I of the periodic table (p.119) and have one s-electron (p.112) in their outer shells (p.111).

alkaline earth metal a metal (↑) which forms an ion (p.123) with an oxidation state (p.135) of +2 only. The oxide is sparingly soluble in water forming a hydroxide (p.48). Alkaline earth metals are chemically reactive (p.62); they are the elements of group II of the periodic table (p.119) and have two s-electrons (p.112) in their outer shell (p.111).

coinage metal a metal (↑) which which does not oxidize (p.70) readily in the air, e.g. copper, silver, gold. The coinage metals are not chemically reactive (p.62); they form positive ions with variable (p.218) oxidation states (p.135); they are all transitional elements (p.121).

coinage metal

copper

silver

gold

iron

lead

zinc

base metals

base metal a metal (↑) which is oxidized (p.70) by heating in air and is acted upon by mineral acids (p.55). Base metals are used for making articles for everyday use. For example, iron, lead, tin, zinc; some are transitional metals, e.g. iron, tin, lead.

halogen (*n*) a non-metal (↑), which forms a negative ion (p.123) with an electrovalency (p.134) of − 1 and is in group VII of the periodic table (p.119). The halogens are fluorine, chlorine, bromine and iodine.

allotropy

three allotropes of carbon
same chemical properties
different physical properties

diamond

graphite

amorphous
carbon

allotropy (*n*) the existence of two or more
different forms of an element without a change
of state (p.9). If the forms are crystalline (p.15)
they are polymorphic (p.92) as well as
allotropic, e.g. sulphur exists in five allotropes
(↓); carbon has two crystalline allotropes.
allotropic (*adj*), **allotrope** (*n*).

dynamic allotropy a kind of allotropy in which the
allotropes are in a dynamic equilibrium (p.150)
with each other, e.g. liquid sulphur has three
allotropes (↓) which show dynamic allotropy.

allotrope (*n*) one of the forms of an element
showing allotropy (↑).

carbon (*n*) a non-metal (p.116) with atomic
number (p.113) 6, relative atomic mass (p.113)
of 12.01 and in group IV of the periodic table
(↓). *See isotope (p.114)*. Carbon is one of the
most important elements as it is a constituent
(p.54) of all living things. It occurs (p.63) in two
crystalline forms, diamond (↓) and graphite
(↓). **carbonic** (*adj*), **carbonaceous** (*adj*).

diamond (*n*) a crystalline form of carbon with a
tetrahedral (p.83) lattice (p.92). It is the
hardest natural substance; it does not conduct
electric current.

graphite (*n*) a crystalline form of carbon with a
hexagonal (p.96) lattice (p.92). It is a soft
substance that marks paper. It is the only non-
metal that conducts electric current.

periodic system if the elements are arranged in order of increasing atomic number (p.113), a periodicity (p.120) is seen in their properties. For example, the elements with atomic numbers 3, 11, 19 are all chemically active metals; atomic numbers 9, 17, 35, 53 are all chemically active non-metals with an electrovalency (p.134) of − 1; atomic numbers 4, 12, 20, 38, 56 are all metals with an oxidation state (p.135) of II. The arithmetical difference between any pair of these numbers is 8 or 18. This suggests a relation (p.232) between the extranuclear (p.113) structure of the electrons and the periodicity of properties. The periodic system is shown by a periodic table (↓).

periodic table the periodic table is shown on p.120 and endpapers. The atomic number (p.113) of each element is shown and the table is arranged in periods (p.120) and groups (↓).

group (*n*) a vertical column of the periodic table (↑). Group I contains the alkali metals (p.117), group II the alkaline earth metals (p.117) and group VII the halogens (p.117). The elements in a group have very similar properties. For metals, chemical reactivity increases from elements with low atomic numbers (p.113) to those with high atomic numbers. For non-metals, chemical reactivity decreases (p.219) from elements with low atomic numbers to those with high atomic numbers. The elements in group I have one s-electron (p.112) in their outer shell (p.111), and those in group II have two s-electrons; elements in group VII have two s-electrons and five p-electrons (p.112).

group (I) alkali metals			
atomic number	relative atomic mass	element	chemical activity
3	6.94	lithium	increases
11	22.98	sodium	
19	39.10	potassium	
37	85.47	rubidium	↓
55	132.91	caesium	
group (VII) halogens			
9	18.99	fluorine	↑
17	35.45	chlorine	
35	79.91	bromine	
53	126.90	iodine	increases

period	group 1	group 2		group 3	group 4	group 5	group 6	group 7	group 8
1	1 H 1.01								2 He 4.00
2	3 Li 6.94	4 Be 9.01		5 B 10.81	6 C 12.01	7 N 14.01	8 O 16.00	9 F 19.00	10 Ne 20.18
3	11 Na 22.99	12 Mg 24.31		13 Al 26.98	14 Si 28.09	15 P 30.97	16 S 32.06	17 Cl 35.45	18 Ar 39.45
4	19 Pk 39.10	20 Ca 40.08	transition elements	31 Ga 69.72	32 Ge 72.59	33 As 74.92	34 Se 78.96	35 Br 79.90	36 Kr 83.80
5	37 Rb 85.47	38 Sr 87.62		49 In 114.82	50 Sn 118.69	51 Sb 121.75	52 Te 127.90	53 I 126.90	54 Xe 131.30
6	55 Cs 132.91	56 Ba 137.33		81 Th 204.37	82 Pb 207.2	83 Bi 208.98	84 Po (209)	85 At (210)	86 Rn (222)
7	87 Fr (223)	88 Ra 226.03							

atomic number
symbol
relative atomic mass

periodic table

period (*n*) a horizontal row of the periodic table
(p.119). It represents the gradual filling of s-
orbitals (p.112) and p-orbitals (p.112), and for
the transitional elements, the d-orbitals (p.112).
The elements in a period change gradually
from a characteristic (p.9) metal on the left of
the period to a characteristic non-metal on the
right. **periodic** (*adj*), **periodicity** (*n*).

periodicity (*n*) the regular occurrence (p.63) of
similar chemical properties with increasing
atomic number (p.113). This is chemical
periodicity.

classify (*v*) to put materials, substances (p.8),
objects, processes (p.157) into classes. In
chemistry, to put elements into groups (p.119)
of the periodic system; to put compounds into
classes such as acids, alkalis, etc.
classification (*n*).

period	transition elements									
4	21 Sc 44.96	22 Ti 47.90	23 V 50.94	24 Cr 52.00	25 Mn 54.94	26 Fe 55.85	27 Co 58.99	28 Ni 58.70	29 Cu 63.55	30 Zn 65.38
5	39 Y 88.91	40 Zr 91.22	41 Nb 92.91	42 Mo 95.94	43 Tc 98.91	44 Ru 101.07	45 Rh 102.91	46 Pd 106.4	47 Ag 107.87	48 Cd 112.41
6	57• La 138.91	72 Hf 178.49	73 Ta 180.95	74 W 183.85	75 Re 186.21	76 Os 190.2	77 Ir 192.22	78 Pt 195.09	79 Au 196.97	80 Hg 200.59
7	89•• Ac 227.03	104 Rf (261)	105 Hn (260)	106 (263)	atomic number symbol relative atomic mass	•Lanthanide elements 57–71 ••Actinide elements 89–92 **transition elements**				

transition element an element in one of three periods (↑), 4, 5 and 6, each period containing 10 transitional elements. The first three periods of the periodic table (p.119) contain no transition elements. The transition elements are formed by the five d-orbitals (p.112) being filled in turn. The extranuclear (p.113) electrons cause the transitional elements to possess variable (p.218) electrovalencies (p.134) and the characteristics (p.9) of metals. These elements form complex ions (p.132) and many coloured compounds.

s-block elements the elements of groups I and II. They have 1 or 2 s-electrons (p.112) in their outer shell (p.111).

p-block elements the elements of groups III to VIII inclusive. They have 1 – 6 p-electrons (p.112) in their outer shell (p.111).

d-block elements the transitional elements. They have s-electrons (p.112) in their outer shell (p.111) and 1 – 10 d-electrons (p.112) in the next inner shell. The number and arrangement of the d-electrons determine the nature of the element.

f-block elements an inner transition series. They have s-electrons in the outer orbitals and fill the f-orbitals, allowing for 14 elements in all.

transition element
extranuclear electrons
of manganese

shell	s-electron	p-electrons	d-electrons
K	2		
L	2	2 2 2 / 6	complete shell
M	2	2 2 2 / 6	1 1 1 1 1 / 5
N	2		

current (*n*) electric charges (p.138) in motion form a current. A flow of electrons is the current of electric charge. Electric current is measured in amperes; it is said to flow from positive to negative.

conduct (*v*) to give direction to the flow of a fluid (p.11), electric current (↑) or heat, e.g. a pipe conducts water; a copper wire conducts an electric current. **conductor** (*n*), **conducting** (*adj*).

conductivity (*n*) a measure of the ability of a solution of given concentration to conduct electric current; its symbol is κ. The conductance of a solution is equal to 1/resistance; its symbol is G. $\kappa = \frac{G}{a}$ where a is a constant (p.106) for a particular electrolytic cell (↓).

electrolyte (*n*) a compound (p.8) which when dissolved in water will conduct (↑) electric current (↑). An electrolyte will also conduct electric current when molten (p.10). Acids, alkalis, inorganic salts (p.46) are generally electrolytes. The electrolyte is decomposed (p.65) by the current.

non-electrolyte (*n*) a compound (p.8) which when dissolved in a solvent (p.86) does not conduct electric current (↑). When molten, a non-electrolyte also does not conduct an electric current. Organic (p.55) compounds generally are non-electrolytes.

electrolytic cell a vessel in which electrolysis (↓) takes place. *See voltameter, p.129.*

electrolysis (*n*) the decomposition (p.65) of an electrolyte (↑) caused by passing an electric current through the solution. **electrolytic** (*adj*).

electrolytic (*adj*) describes any process (p.157) connected with electrolysis (↑).

electrode (*n*) a piece of material that conducts (↑) an electric current (↑) used in an electrolytic cell (↑) or voltameter (p.129) or simple cell (p.129). Electric current enters and leaves the solution of the electrolyte through the electrodes; the electrodes are connected (p.24) to a source (p.138) of electric current in an electrolytic cell or voltameter.

electrolyte

electrolytes conduct electric current

fused electrolyte

heat

non-electrolytes do not conduct electric current

non-electrolyte

electrolytic cell

anode ⊕ | ⊖ cathode

electrode — electrode

electrolyte in solution

ammeter

ions and ionization

ammeter (*n*) an instrument (p.23) for measuring the strength of an electric current (↑) in amperes.

anode (*n*) a positive electrode (↑); the electrode at which electric current is said to enter an electrolytic cell; the electrode from which electric current leaves a simple cell. **anodic** (*adj*).

cathode (*n*) a negative electrode (↑); the electrode from which current is said to leave an electrolytic cell; the electrode at which electric current flows into a simple cell after conduction by conductors.

electrolyze (*v*) to pass an electric current (↑) through a solution of an electrolyte (↑), or a molten (p.10) electrolyte, in order to decompose it, e.g. copper metal is obtained by electrolyzing a solution of copper (II) sulphate. **electrolysis** (*n*), **electrolytic** (*adj*).

ion (*n*) an ion is formed by an atom gaining or losing one or more valency (p.133) electrons (p.110). An atom, or a group of atoms in an ion, is neutral; the loss or gain of electrons results in an electric charge on the ion. Loss of electrons produces a positive charge; gain of electrons produces a negative charge. The magnitude of the charge depends upon the number of electrons concerned. The process of forming ions is ionization (↓). **ionize** (*v*).

ionization (*n*) the process of forming ions (↑). Ions are formed: (1) when energy is supplied to atoms; the energy can be heat or radiation; a high voltage (p.126) will also cause ions to be formed. (2) the attraction for electrons by atoms of some elements, e.g. a chlorine atom attracts one electron from a sodium atom. Crystals of electrolytes (↑) have a lattice consisting of ions; dissolving the crystals in water separates the ions, but this is not ionization. Hydrogen chloride is a covalent (p.136) compound, when it dissolves in water ionization takes place because the chlorine atom attracts an electron from the hydrogen atom. **ionize** (*v*), **ionized** (*adj*), **ionizing** (*adj*).

attract (v) to pull one object towards another object, e.g. a positive ion (p.123) attracts a negative ion towards itself; positive and negative electric charges attract each other. **attraction** (n), **attractive** (adj).

repel (v) to push one object away from another object, e.g. two positive ions (p.123) repel each other; two negative ions repel each other. **repulsion** (n), **repulsive** (adj).

repulsion (n) the force which causes objects to repel each other, e.g. a force of repulsion exists (p.213) between two positive electric charges and between two negative electric charges. **repulsive** (adj).

ionic theory the theory which explains the behaviour of electrolytes (p.122). It accounts for electrolysis (p.122) and the action of simple cells (p.129) by the existence (p.213) of ions (p.123) in all electrolytes.

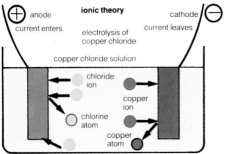

discharge (v) to give up an electric charge. When an ion (p.123) comes into contact (p.217) with an electrode (p.122), the ion either receives or gives electrons and becomes a neutral atom; the ion has given its charge to the electrode and has become discharged. In the case of metals, it may be deposited on the electrode, in the case of non-metals, it is generally liberated (p.69) as a gas.

strength (n) a measure of the degree (p.227) of ionization (p.123) of an electrolyte (p.122) in a solution; also a measure of the characteristic reactivity (p.62) of acids (p.45) and alkalis (p.45).

strong electrolyte an electrolyte (p.122) which is
completely ionized (p.123) even in concentrated
solutions. Strong acids (p.45) and strong alkalis
(p.45) are strong electrolytes. Most inorganic
salts are strong electrolytes, but *see hydrolysis*[1]
(p.66). Strong electrolytes are good conductors
(p.122) of electric current.

weak electrolyte an electrolyte which is only
partly ionized (p.123) in water, or other ionizing
solvents (p.86). At great dilutions the ionization
becomes almost complete. Solutions of weak
electrolytes are poor conductors (p.122) of electric current (p.122). Weak acids (p.45) and weak
alkalis (p.45) are weak electrolytes. Examples of
weak electrolytes are: ethanoic acid (and other
organic acids); ammonia solution (a weak alkali).

anion (*n*) an ion which carries a negative charge; it
is attracted to the anode (p.123) in electrolysis
(p.122). Examples of anions are: chloride ion Cl^-;
sulphate ion SO_4^{2-}; zincate (II) ion ZnO_2^{2-}.

cation (*n*) an ion which carries a positive charge;
it is attracted to the cathode (p.123) in
electrolysis (p.122). Examples of cations are:
copper (II) Cu^{2+}; sodium Na^+; iron (III) Fe^{3+}

voltage (*n*) a measurement in volts of the electrical force, or pressure, that drives an electric current (p.122) through an electrolyte (p.122) or conductor (p.122).

voltmeter (*n*) an instrument (p.23) that measures voltage (↑) in volts.

decomposition voltage the smallest voltage (↑) that will cause decomposition (p.65) of an electrolyte (p.122) by electrolysis (p.122) using platinum or other electrodes (↓). This voltage is necessary to overcome (p.213) the effects of polarization (↓) and overvoltage (↓). With some electrodes and electrolytes it is 0 volts.

voltmeter

decomposition voltage

platinum electrode an electrode consisting of platinum; such electrodes produce polarization (↓) and overvoltage (↓). *Platinized electrodes* consist of platinum electrodes with a layer of finely divided platinum called *platinum black;* such electrodes do not produce polarization.

polarization

polarization (*n*) the production of hydrogen and oxygen gases which collect on an electrode, changing the nature of the electrode. In electrolysis (p.122) a primary cell (p.129) is formed with hydrogen and oxygen electrodes; this produces an *electromotive force* (p.129) acting against the electrolyzing current and causes a decomposition voltage (↑) to be necessary. In primary cells, the electromotive force produced by polarization acts against the electromotive force of the cell and reduces its voltage (↑).

overvoltage (*n*) the further voltage necessary to release (p.69) a gas at an electrode above its *electrode potential* (p.128). The magnitude of the overvoltage depends on the nature of the electrode and the gas being discharged (p.124). The cause of overvoltage has not been explained.

electrodeposit (*n*) a coat (↓) of metal on an electrode (p.122) produced during electrolysis by the discharge (p.124) of ions of the metal at the electrode. **electrodeposition** (*n*).

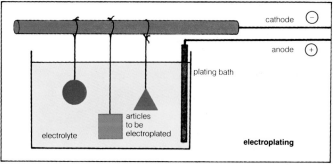

cathode ⊖

anode ⊕

plating bath

articles
to be
electroplated

electrolyte

electroplating

electroplating (*n*) the process (p.157) of putting an electrodeposit (↑) on an article. In electroplating with silver, a metal article is made the cathode (p.123) of a *plating bath* and a solution of a silver salt is used as the electrolyte. A thin coat of silver is formed on the article.

plate (*v*) to carry out electroplating (↑), especially with silver.

coat (*n*) a thin outer layer on an object. **coat** (*v*).

anodic (*adj*) describes anything happening at an anode (p.123) during electrolysis (p.122).

cathodic (*adj*) describes anything happening at a cathode (p.123) during electrolysis (p.122).

cathodic reduction a process (p.157) at a cathode (p.123) during electrolysis, in which a positive ion (p.123) gains one or more electrons from the cathode, e.g. hydrogen ions gain one electron to become hydrogen atoms, the ion is reduced (p.219) to the atom and discharged (p.124).

anodic oxidation a process (p.157) at an anode (p.123) during electrolysis (p.122), in which a negative ion (p.123) loses one or more electrons to the anode. A metal anode may go into solution, as its atoms can lose one or more electrons to become positive ions, e.g. negative ions:
$$2\,Cl^- \rightarrow Cl_2 + 2e \text{ (chlorine discharged)}$$
atoms:
$$Cu \rightarrow Cu^{2+} + 2e \text{ (electrode goes into solution).}$$

anodize (*v*) to form a layer of oxide (p.48) on a metal by electrolysis (p.122). The metal is made the anode (p.123) of an electrolytic cell (p.122), a suitable electrolyte is used and oxygen is discharged (p.124) at the anode forming a coat of the oxide of the metal. With aluminium, dyes (p.162) can be absorbed in the oxide, producing coloured surfaces. **anodizing** (*n*).

electrochemical (*adj*) describes any effect concerned with the electrical properties of solutions (p.86) and the ions (p.123) in solution.

electrochemical equivalent the mass in grams of an element liberated (p.69) during electrolysis (p.122) by 1 coulomb of electric charge. (1 coulomb of charge is the quantity of charge from a current of 1 ampere flowing for 1 second).

electrode potential when a metal is put in a solution containing ions (p.123) of the metal, a voltage (p.126) difference forms between the metal and the solution; this is an electrode potential. An electrode potential is also formed between a metal and any electrolyte. Two metals will have two electrode potentials and the difference between these potentials produces the electromotive force (↓) of a primary cell.

ammeter battery

Ag 1.08 g

electrochemical equivalent

965 coulomb

Cu 0.3175 g

gases evolved

Hofmann voltameter

electrolyte

graduated tube

platinum electrodes

H_2 0.01 g O_2 0.16 g

Hofmann voltameter

copper voltameter

ammeter

battery

copper electrodes

copper (II) sulphate solution

series[1] a way of connecting (p.24) parts of an electric circuit so that the current flows through each part in turn. The parts are said to be connected in series.

parallel (*n*) a way of connecting (p.24) parts of an electric circuit so that the current divides to go through each part at the same time. The parts are said to be connected in parallel.

voltameter (*n*) an apparatus for determining the electrochemical equivalent (↑) of an element. The electrodes are large metal plates, weighed before and after passing a measured electric current for a measured time. A Hoffman voltameter is used to determine the electrochemical equivalent of gases. *See electrolytic cell (p.122).*

faraday (*n*) the quantity of electric current that liberates (p.69) or forms 1 mole of monovalent (p.137) ions. Its value is 96 487 coulombs. *See electrochemical equivalent (↑).* One faraday of electric charge contains one mole of electrons.

inactive electrode an electrode which will not give ions (p.123) to the electrolyte (p.122) in an electrolytic cell (p.122), e.g. carbon and platinum are inactive electrodes.

active electrode an electrode which can give ions (p.123) to the electrolyte (p.122) in an electrolytic cell (p.122) if it is the anode. The cathode is not so important. For example, copper forms an active anode and may produce copper ions during electrolysis (p.122).

primary cell

voltmeter

1.1 v

(negative)

(positive)

zinc electrode

copper electrode

dilute sulphuric acid

(electrolyte)

primary cell a source (p.138) of electric current formed by putting two metals into an electrolyte. Also called a voltaic cell. The source of the electrical energy is the *electrode potentials* (↑) between each of the electrodes and the electrolyte in the cell, e.g. a simple voltaic cell has an electrode of zinc and one of copper in an electrolyte of dilute sulphuric acid; it has an electromotive force of 1.1 volt.

electromotive force the electrical force of a primary cell (↑) which drives current through a connected circuit. The symbol for electromotive force is *E*, and the unit for it is the volt. The electromotive force is produced by the two *electrode potentials* (↑) of the cell. (*abbr.*) **e.m.f.**

hydrogen electrode a platinum electrode
(p.126) is platinized with platinum black and
hydrogen gas is put in contact with the
electrode at standard pressure (p.102). This
electrode is placed in a 1 M (p.88) solution of
hydrogen ions (p.123) at 25°C to form a
hydrogen electrode. The *standard electrode
potential* (↓) of the hydrogen electrode is fixed
at 0 volts.

standard electrode potential the electrode
potential (p.128) of a metal, or a gas, in contact
with 1 M (p.88) solution of its ions at 25°C; the
magnitude is determined (p.222) against the
standard hydrogen electrode (↑) of 0 volts.

electrochemical series a list of elements in
order of their standard electrode potentials (↑)
using a scale with the hydrogen electrode (↑)
as zero. In this series, a metal such as zinc has
a negative electrode potential as in a primary
cell with a zinc and a hydrogen electrode, the
zinc would be the cathode. *See below for the
electrochemical series*. Similarly, in a primary
cell with a chlorine and a hydrogen electrode,
the chlorine would be the anode and so
chlorine has a positive electrode potential.

**standard
electrode
potential**

electrochemical series		
element	electrode potential	ionic reaction
potassium	−2.92 v	$K^+ + e \rightarrow K$
calcium	−2.87 v	$Ca^{2+} + 2e \rightarrow Ca$
sodium	−2.71 v	$Na^+ + e \rightarrow Na$
magnesium	−2.38 v	$Mg^{2+} + 2e \rightarrow Mg$
aluminium	−1.66 v	$Al^{3+} + 3e \rightarrow Al$
zinc	−0.76 v	$Zn^{2+} + 2e \rightarrow Zn$
iron	−0.44 v	$Fe^{2+} + 2e \rightarrow Fe$
tin	−0.14 v	$Sn^{2+} + 2e \rightarrow Sn$
lead	−0.1 v	$Pb^{2+} + 2e \rightarrow Pb$
hydrogen	0 v	$H^+ + e \rightarrow H$
copper	+0.34 v	$Cu^{2+} + 2e \rightarrow Cu$
silver	+0.8 v	$Ag^+ + e \rightarrow Ag$

redox potential a metal acts as a reducing agent
(p.71) and a non-metal, such as chlorine, acts
as an oxidizing agent (p.71). The electrochemical series (↑) shows the relative (p.232)
strengths of these agents. By using the apparatus
of redox half cells, other ions can be used to
determine (p.222) their electrode potential and
these potentials are called redox potentials.
Redox potentials show whether an agent is able,
or not, to reduce or to oxidize an ion. For example,
the redox potential of chlorine to chlorine ions
is + 1.36 V; the redox potential of $(Cr_2O_7)^{2-}$ is
1.33 V and of $(MnO_4)^-$ is + 1.52 V. Hence
potassium manganate (VII) will oxidize chlorine
ions to chlorine, but potassium dichromate (VI)
will not usually oxidize chlorine ions.

redox series a list of redox reagents, which
include the electrochemical series, in order of
their redox potentials, measured at 25°C, in
1 M (p.88) solutions, forms the redox series.
Hydrogen is given a redox potential of zero.

redox series	
$Sn^{4+} + 2e \rightarrow Sn^{2+}$	+0.15 v
$Cu^{2+} + 2e \rightarrow Cu$	+0.34 v
$2H_2O + O_2 + 4e \rightarrow 4OH^-$	+0.44 v
$Fe^{3+} + e \rightarrow Fe^{2+}$	+0.77 v
$Cr_2O_7{}^{2-} + 14H_2{}^+ + 6e \rightarrow 2Cr^{3+} + 7H_2O$	+1.33 v
$Cl_2 + 2e \rightarrow 2Cl^-$	+1.36 v
$MnO_4{}^- + 8H^+ + 5e \rightarrow Mn^{2+} + 4H_2O$	+ 1.52 v

an aqua-ion with tetrahedral structure **2**+

H

Cu

O

tetraquo copper (II)

complex ion

complex ion an ion (p.123) which has molecules (p.77) joined to it by coordinate bonds (p.136). A complex cation (p.125) has molecules or atoms such as water, a halogen, or cyanide (CN) joined by coordinate bonds to the metal ion or to hydrogen and carries a positive charge, e.g. tetrammine copper (II) ion $[Cu(NH_4)_4]^{2+}$, tetrachlorodiaquo chromium (III) $[CrCl_4(H_2O)_2]^{3+}$. A complex anion (p.125) has molecules or atoms such as water, a halogen or cyanide (CN) joined by coordinate bonds to a central metal atom and carries a negative charge, e.g. hexacyanoferrate (II) $[Fe(CN)_6]^{4-}$ trichlorocuprate (I) $[CuCl_3]^{2-}$. The central metal atom or ion in complex ions is usually a transitional element (p.121).

aqua-ion (*n*) a complex ion (↑) with water molecules joined to a metal ion by coordinate bonds, e.g. tetraquo copper (II) $[Cu(H_2O)_4]^{2+}$

hydroxide ion the ion formed from a water molecule by the removal of a hydrogen ion. It has a negative charge; it is discharged at an anode to produce oxygen:
$$4OH^- \rightarrow 2H_2O + O_2 + 4e$$

transfer (*v*) to move objects, materials, or energy from one place to another, without saying how they are moved, e.g. to transfer an electron from an ion to an electrode; to transfer a precipitate from a filter paper to a beaker. **transference** (*n*), **transferable** (*adj*).

valency electron

model of a
carbon – 12 atom

valency shell

model of an electron
pair in an orbital

shared electron

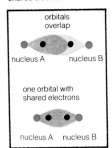

bond (*n*) in chemistry, a force which holds together two atoms, two ions, two molecules, or a combination of these. Bonds are broken and formed in chemical change. Energy is needed to break bonds; it is released or absorbed in forming bonds.

valency (*n*) the ability of an atom, and hence an element, to form bonds (↑), e.g. carbon, valency 4, can form 4 bonds. *Valency* has now been replaced (p.68) by *electrovalency* (p.134) and *covalency* (p.136).

valency electron an electron (p.110) which takes part in chemical bonds (↑).

valency shell the outside shell (p.111) of an atom which contains valency electrons (↑).

electron pair an orbital (p.111) can hold two electrons (p.110) and the orbital is then full. An orbital in a valency shell (↑) can lose electrons or gain electrons to form an ion (p.123). An orbital can gain an electron by sharing an electron from another orbital; this forms a covalent bond. Full orbitals are stable (p.74).

lone pair an orbital (p.111) in a valency shell (↑) containing two electrons not involved in bonding in a molecule. A lone pair can take part in a coordinate bond (p.136), e.g. nitrogen has 5 electrons in its valency shell, 2 in one orbital forming a lone pair and 3, each in one orbital, available (p.85) to form bonds (↑).

octet (*n*) the eight electrons which fill the s- and p-orbitals (p.112) of the valency shell (↑) of an atom. Except for hydrogen, helium and the transition elements (p.121), the valency shell of an atom has one s- and three p-orbitals. When these orbitals are full a very stable structure is formed as in the noble gases. Atoms with a valency shell of s- and p-orbitals tend to form ions with no electrons or an octet in the outer shell, giving a stable atomic structure.

shared electron two orbitals (p.111), each with one electron (p.110), overlap (p.218). Each orbital shares its electron with the other orbital. In this way each orbital has its own electron and a shared electron, so each orbital can be considered to have two electrons.

ionic bond a kind of chemical bond (p.133) formed between ions (p.123). It is usually formed between the ions of a metal and a non-metal (p.116). The metal atom loses one or more electrons (p.110) to form a positive ion and the non-metal atom gains one or more electrons to form a negative ion, e.g. sodium burns in chlorine to form sodium chloride; a sodium atom loses 1 electron to form a sodium ion and a chlorine atom gains 1 electron to form a chloride ion. In solution the ions are (solvated (p.90) and) free to move; in a crystal (p.91) the ions are held together by the attraction (p.124) of a positive charge for a negative charge and the ions form a giant structure (p.94). Attraction between opposite charges forms an ionic bond.

electrovalent bond = ionic bond (↑).

electrovalency (*n*) the number of ionic bonds (↑) an atom can form. This is equal to the number of electrons (p.110) which it can lose or gain in its valency shell (p.133) to form an ion. The electrovalency of an atom is also the electrovalency of the element. For example, a magnesium atom has two electrons in its valency shell, both can be removed (p.215) to form an ion with 2 positive charges, so magnesium has an electrovalency of + 2 (this shows the ion is positive). A chlorine atom has 7 electrons in its valency shell and it can gain one electron to form an ion with 1 negative charge, so chlorine has an electrovalency of −1 (this shows the ion is negative). In both cases (magnesium and chlorine) the atoms form ions with a stable (p.74) octet (p.133) in the outside shell.

oxidation state this is determined by the charge on an ion (p.123) of an element (p.116). Copper, with an electrovalency ($\uparrow$) of $+2$, forms an ion Cu^{2+}; the oxidation state of copper is $+2$. Copper with an electrovalency of $+1$ forms an ion Cu^+, the oxidation state is $+1$. Similarly, the oxidation state of Cl^- is -1 and of S^{2-} is -2. The oxidation state indicates (p.38) the number of electrons that have been removed (p.215) from, or added to, an atom to form an ion, e.g. if the oxidation state is $+3$, then three electrons have been removed from a neutral atom. *See oxidation number (p.78).*

copper atom
oxidation state O

oxidation state

copper (II) ion
oxidation state $+2$

copper (I) ion
oxidation state $+1$

energy (*n*) energy can be used to do work. Chemical energy is available (p.85) in a material, or substance (p.8), from the energy stored in its chemical bonds (p.133). This energy is set free when the bonds are broken or new bonds made. Chemical energy is often transformed (p.144) into heat energy in a chemical reaction, e.g. in combustion (p.58); the heat energy can be used to do work as in a motor car engine. Chemical energy is transformed into electrical energy in a primary cell (p.129); the electrical energy can be used to do work. **energize** (*v*).

covalent bond a kind of chemical bond (p.133) formed between atoms (p.110) by shared electrons (p.133). In simple molecules, covalent bonds are usually formed by p-electrons (p.112) in p-orbitals (p.112). These orbitals are directed in space, so covalent bonds are directed in space and account for the shape of molecules. After an atom has formed covalent bonds in simple molecules, it has a complete octet (p.133) of electrons. Compounds with covalent bonds form molecular crystals (p.94) or very hard giant structures (p.94).

covalency (n) the number of covalent bonds (↑) an atom (p.110) can form. This is equal to the number of electrons (p.110) which are single electrons in orbitals (p.111); such electrons are available (p.85) to become shared electrons (p.133), e.g. oxygen has a covalency of 2 because it has two unpaired electrons (p.110) able to form two covalent bonds. **covalent** (adj).

bond energy the energy given out, or taken in, when a covalent bond is formed between two free atoms. It is also the energy needed to break the bond; and is thus a measure of the strength of the bond, e.g. the bond energy of the C–H bond is 415 kJ mol^{-1}

coordinate bond a kind of covalent bond (↑) formed between two atoms when one atom gives both electrons to form the shared electrons (p.133) of the bond. The electrons that are given are a lone pair (p.133). The ammonium ion is formed by a coordinate bond between the nitrogen atom of ammonia and a hydrogen ion, *see diagram.*

dative bond another name for coordinate bond (↑), not used now.

dative covalent bond another name for coordinate bond (↑), not used now.

semipolar bond another name for coordinate bond (↑), not used now.

donor (n) the atom which gives a lone pair in a coordinate bond (↑).

acceptor (n) the atom which receives a lone pair in a coordinate bond (↑).

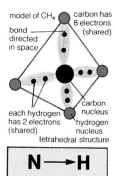

model of CH$_4$
carbon has 8 electrons (shared)
bond directed in space
carbon nucleus
each hydrogen has 2 electrons (shared)
hydrogen nucleus
tetrahedral structure

co-ordinate bond

diagram of CH$_4$

H
H • C • H
H

covalent bond

donor
acceptor
donor
lone pair
acceptor
hydrogen ion

ammonium ion

van de Waals' bonds

hexagonal structure

polar

polar molecule of water (H_2O)

metallic bond a kind of chemical bond (p.133) found in metal crystals (p.95). The crystal consists (p.55) of positive ions of the metal in a lattice (p.92) with the valency electrons (p.133) free to move between the metal ions. The moving electrons form the metallic bond.

van der Waals' bond a weak chemical bond (p.133) which holds molecules together. It is an attractive (p.124) force arising from the movement of electrons in atoms and is ten to twenty times weaker than the attractive force between ions. *See molecular crystals (p.94).* Graphite is an example of molecules held by van der Waals' bonds. The carbon atoms form hexagons in layers, held by strong covalent bonds (↑). The layers are held together by van der Waals' bonds so that the layers slip over each other, explaining the softness of graphite.

polar (*adj*) describes a molecule (p.77) in which the covalent bonds (↑) have the pair of electrons forming the bond nearer to one atom than another. This makes one atom slightly negative and the other atom slightly positive, each with a fraction of the charge on an electron (shown by $\delta +$ or $\delta -$), e.g. the water molecule is polar, *see diagram*. If the substance is liquid, a polar solvent (p.86) is formed. Polar solvents dissolve ionic (p.123) compounds.

non-polar (*adj*) describes a molecule (p.77) which is not polar (↑). If the substance is liquid, a non-polar solvent (p.86) is formed. Non-polar solvents dissolve organic (p.55) compounds, which are generally non-polar themselves.

monovalent (*adj*) describes an element with a valency of 1; it is better described as monoelectrovalent or monocovalent.

divalent (*adj*) describes an element with a valency of 2; it is better described by the electrovalency or covalency.

tervalent (*adj*) describes an element with a valency of 3; it is better described by the electrovalency or covalency.

radioactivity (*n*) the property of spontaneous
(p.75) nuclear (p.110) changes in which
energy is released as radiation (↓) and a new
nucleus is formed with a different number of
protons and, in some cases, a different
number of neutrons. **radioactive** (*adj*).

radiation (*n*) (1) a process in which energy is
passed on in the form of a wave motion, with
electromagnetic waves. Examples of radiation
are light, ultra-violet light, X-rays, radio waves.
(2) a form of energy, travelling in straight lines
and causing ionization (p.123) of any material
through which it passes. **radiant** (*adj*).

radioactive (*adj*) describes an element, or one
of its compounds, which has the property of
radioactivity (↑).

natural radioactivity the radioactivity (↑) of
elements with atomic numbers (p.113) greater
than 83. Such elements have radioactive
compounds which occur (p.63) naturally (p.19).

artificial radioactivity the radioactivity of
elements which have been made radioactive
by bombardment (p.143) with particles (p.13)
of high energy.

radium – 224

radioactivity ↓

radiation + radon – 220

radioactivity 1

radioactive element

source emits rays

source

rays of radiation

radioactivity 2

charge (*n*) no description of charge can be
made other than it exists (p.213) in two forms,
positive and *negative*. It is a property of subatomic particles (p.13); an electric current is a
flow of charge. **charged** (*adj*).

ray (*n*) (1) a line which shows the direction of
radiation (↑). (2) a line of radiant (↑) energy.
(3) a stream of charged (↑) particles (p.13).

source (*n*) a place from which something, such
as radiation (↑), or an ore (p.154), has come,
e.g. a radioactive element is a source of radiation; a primary cell is a source of electric current.

emit (*v*) to give out radiation, gas, odour, sound,
from a source (↑), e.g. a bell emits sound when
hit; a radioactive element emits radiation,
emission (*n*), **emissive** (*adj*), **emitter** (*n*).

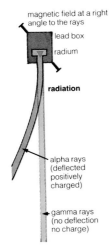

magnetic field at a right angle to the rays

lead box

radium

radiation

alpha rays (deflected positively charged)

gamma rays (no deflection no charge)

a radioactive source emits alpha rays or alpha rays and gamma rays

Becquerel rays the rays (↑) emitted by a radioactive source. The kind of ray is not described.

alpha ray a ray (↑) consisting (p.55) of alpha particles (↓).

beta ray a ray (↑) consisting (p.55) of beta particles (↓).

gamma ray a ray (↑) of gamma radiation (↓).

alpha particle the nucleus of a helium atom consisting (p.55) of two protons and two neutrons; it has a relative atomic mass of 4.0029 and a charge of + 2. An alpha particle has a range (p.140) in air of 7 cm and ionizes (p.123) gases through which it passes. Alpha particles cause scintillation (p.140) when they hit fluorescent (p.140) surfaces.

α**-particle** another way of writing alpha particle (↑).

beta particle an electron (p.110) or positron (p.110) emitted (↑) from the nucleus of a radioactive atom; it has a very high speed, up to 99% of the speed of light. Loss of an electron results in the nucleus increasing its number of protons by 1 and decreasing its neutrons by 1. A beta particle has a range (p.140) in air of 750 cm and can penetrate (p.144) thin pieces of metal foil. It ionizes (p.123) gases through which it passes but the effect is weaker than that of alpha particles (↑).

ß-particle another way of writing beta particle.

gamma radiation radiation (↑) emitted by radioactive (↑) elements; it has wavelengths shorter than those of X-rays. Gamma radiation is emitted with either alpha or beta radiation. The radiation is very penetrating (p.144) with a range of 150 mm in lead; it carries no charge, so is not deflected by electric or magnetic fields. The wavelength of a radiation is a characteristic (p.9) of the nucleus emitting it. Gamma radiation has a very weak ionizing (p.123) effect.

γ**-radiation** a way of writing gamma radiation.

beta rays (deflected, negatively charged)

proactinium

lead box

gamma rays (no deflection, no charge)

radiation

magnetic field at a right angle to the rays

a radioactive source emits beta rays or beta rays and gamma rays

alpha emission the emission of an alpha particle
(p.139) from the nucleus of a radioactive
(p.138) atom. The resulting nucleus has an
atomic number (p.113) which is 2 less and a
mass number (p.113) which is 4 less, e.g.
$$^{220}_{86}Rn \rightarrow {}^{216}_{84}Po + {}^{4}_{2}He \text{ ($\propto$-particle)}$$
Alpha emission may be accompanied (p.213)
by gamma radiation; it is never accompanied
by beta emission (↓).

beta emission the emission of a beta particle
(p.139) from the nucleus of a radioactive (p.138)
atom. The resulting nucleus has an atomic
number (p.113) which is 1 more, but the mass
number (p.113) remains the same, e.g
$$^{228}_{88}Ra \rightarrow {}^{228}_{89}Ac + {}^{0}_{-1}e$$
In the nucleus, a neutron change produces a
proton and an electron; the proton remains
and the electron is emitted. Beta emission may
be accompanied (p.213) by gamma radiation;
it is never accompanied by alpha emission (↑).

alpha emission

beta emission

range (n) (1) the greatest distance an object can
travel when it has been given energy, e.g. the
range of an aeroplane on a tank full of fuel
(p.160). (2) a set of values between an upper
and lower limit, e.g. the temperature range
between 0°C and 100°C. **range** (v).

scintillation (n) very short, bright, quick flashes
of light. **scintillate** (v).

fluorescent (adj) a material which emits light
when radiation, of a shorter wavelength than
light, falls on the material. **fluorescence** (n).

spinthariscope (n) a device consisting of a surface coated (p.127) with zinc sulphide. When
alpha particles hit the surface, it scintillates (↑).
The number of scintillations can be counted,
giving the number of alpha particles emitted by
a radioactive source (p.138).

spinthariscope

protective cover

mica window

anode wire

wire gauze

inert gas at 10 cm Hg pressure

metal cathode

Geiger counter

v cm³ $\dfrac{v}{2}$ cm³ $\dfrac{v}{4}$ cm³

radon gas

mercury

3.8 days   3.8 days

radioactive decay of radon
half-life period of 3.8 days

half-life

Geiger counter a device for the detection
(p.225) of ionizing (p.123) radiations. It consists
(p.55) of a metal tube, with a window, and a
wire electrode (p.122) in the centre of the tube.
The tube is filled with argon (an inert gas) at a
low pressure. There is a high electric potential
between the electrode and the tube, with the
electrode made positive. When radiation enters
through the window, the argon atoms are
ionized and the electrode is discharged, giving
a pulse of current, which can be detected.

radioactive disintegration the disintegration
(p.65) of the nucleus of a radioactive element
into two parts. The disintegration usually
results in the formation of another nucleus and
an alpha or beta particle.

radioactive decay the decrease, with time, of
radioactivity (p.138) from a specimen of a
radioactive element.

half-life (n) the time taken for one half of the atoms
in a specimen of a radioactive element to disintegrate (↑). It is an important characteristic (p.9)
of a radioactive element, and is a constant for
that element, e.g. examples of half-lives are:
uranium-238, 4.5×10^9 years; radium-226, 1620
years; radium-221, 30 seconds. The half-life
period is independent of temperature, pressure, concentration and nature of the material.

neutron   92 p 143 n   56 p 88 n   barium – 144

neutrons

36 p 54 n

uranium – 235   krypton – 90

nuclear fission

nuclear fission the disintegration of a nucleus
into smaller nuclei, brought about by bombardment (p.143) with neutrons, e.g. uranium-238
bombarded by low energy neutrons splits into
barium-144 and krypton-90, releasing two
neutrons: $^{235}_{92}U + {}^{1}_{0}n \rightarrow {}^{144}_{56}Ba + {}^{90}_{36}Kr + 2{}^{1}_{0}n$
In this process, energy is released; 1 kg of
uranium by nuclear fission produces
approximately 10^{10}kJ.

fissile (*adj*) describes any element, or
substance, which can undergo fission,
particularly nuclear fission (↑).

nuclear fusion the combination of two nuclei of small mass number to produce one nucleus of a higher mass number; the nuclear reaction takes place only at very high temperatures (about 10^8 K). A large quantity of energy is released, e.g. $^2_1H + {}^1_1H \rightarrow {}^3_2He + 5 \times 10^5$ MJ. Deuterium bombarded with protons forms helium-3, releasing large quantities of energy.

curie (*n*) the unit for the rate of radioactive disintegration (p.141). 1 curie = 3.7×10^{10} atoms disintegrating per second. 2996 kg uranium-238 produce 3.7×10^{10} atomic disintegrations per second, hence this mass of uranium has an activity of 1 curie. The safe limit of radioactivity for human beings is estimated to be 10 microcuries. The symbol for curie is Ci.

radioactive series one of three groups of naturally occurring radioactive (p.138) elements, each series named after the element with which the series starts. The uranium series begins with uranium-238 and passes, by 15 nuclear changes, to lead-206, which is stable (p.74). The thorium series begins with thorium-232 and passes by 11 nuclear changes to lead-208, which is stable. The actinium series starts with protoactinium-231 and passes through 9 nuclear changes to lead-207, which is stable.

disintegration series alternative name for radioactive series (↑).

Fajans and Soddy law the emission of an alpha particle during radioactive change produces an element two places to the left in the periodic table (p.119). The emission of a beta particle, however, produces an element one place to the right in the periodic table.

nuclide (*n*) an atomic species which is defined by the number of protons and neutrons in the nucleus, and by its kind of radioactive decay and the half-life (p.141) of the rate of decay, e.g. radium-221 has 88 protons and 133 neutrons in its nucleus, it decays by alpha-emission with a half-life of 30 seconds. Radium-223 has 88 protons and 135 neutrons in its nucleus, it decays by alpha-emission accompanied (p.213) by gamma-radiation and has a half-life of 11.7 days.

hydrogen – 1   hydrogen – 2 deuterium

helium – 3

+ energy

atomic fusion

radioactive series

$^{231}_{91}$ Pa

↓ $-\alpha$

$^{227}_{89}$ Ac

↓ $-\beta$

$^{227}_{90}$ Th

↓ $-\alpha$

$^{223}_{88}$ Ra

↓ $-\alpha$

$^{219}_{86}$ Rn

↓ $-\alpha$

$^{215}_{84}$ Po

↓ $-\alpha$

$^{211}_{82}$ Pb

↓ $-\beta$

$^{211}_{83}$ Bi

$-\alpha$ $-\beta$

$^{207}_{81}$ Tl   Po $^{211}_{84}$

$-\beta$ $-\alpha$

$^{207}_{82}$ Pb

actinium series

chamber (*n*) a hollow space surrounded by walls, usually with a small entrance.
track (*n*) visible (p.42) signs that an object has travelled along a path. The signs are continuous.

camera

felt soaked in alcohol

radioactive source

region of supersaturated vapour·

strong light

metal plate

transparent plastic

solid carbon dioxide

cloud chamber

sponge

base

cloud chamber a round chamber (↑), made of transparent (p.16) plastic which has a firm base. Above the base, solid carbon dioxide is supported by a sponge. A metal plate rests on the solid carbon dioxide. Round the top of the chamber is a strip of felt, soaked in alcohol (ethanol). The ethanol vaporizes (p.11) and saturates dust-free air in the chamber. The metal plate is very cold and forms a region of supersaturated vapour, just above the plate. A radioactive source produces radiation which ionizes (p.123) the molecules in the air. Drops of liquid condense on the ions, forming a track (↑) which is photographed by a camera, using strong light, as shown in the diagram. Condensation takes place only in the region of supersaturated vapour. Cloud chambers are used to study radiation from radioactive sources and nuclear reactions (p.144).
bombard (*v*) to hit a large object repeatedly with many small objects, each small object possessing considerable energy, **bombardment** (*n*).

tracks of alpha particles in a cloud chamber

penetrate (v) to go through an outer cover, using force or energy, or similarly, to reach inside a solid material, e.g. gamma rays penetrate steel; a neutron penetrates an atomic nucleus. **penetration** (n), **penetrating** (adj).

radio opaque (adj) describes a material which does not allow radiation (p.138) to pass through it because the radiation is absorbed. Radio opaqueness depends on the density of the material and on the wavelength of the radiation. The shorter the wavelength, the greater the penetration (↑). **radio opaqueness** (n).

transform (v) to change one form of energy into another form. **transformation** (n).

radiology (n) the study of X-rays and radioactivity (p.138). In particular, it is the use of radiations (p.138) in medical science for curing diseases. **radiologist** (n), **radiological** (adj).

nuclear reaction a reaction in which changes in the nucleus of an atom take place. Nuclear changes are caused by the bombardment (p.143) of a nucleus with subatomic (p.110) and other particles, e.g.

$$^{63}_{29}Cu + {}^{1}_{1}H \rightarrow {}^{63}_{30}Zn + {}^{1}_{0}n$$

in which copper-63 is bombarded with protons and forms zinc-63 and neutrons. Aluminium-27 bombarded with alpha particles forms phosphorus-30 and neutrons:

$$^{27}_{13}Al + {}^{4}_{2}He \rightarrow {}^{30}_{15}p + {}^{1}_{0}n$$

The products of nuclear reactions are generally radioactive isotopes (p.114) of the element.

a nail penetrates a block of wood to a depth of 3 cm

nuclear reactions

mass spectrograph

mass spectrograph

mass spectrograph of
magnesium

mass spectrograph a piece of apparatus for
determining (p.222) the mass number of isotopes
(p.114). An ion (p.123) source, *see diagram*,
produces positive ions of an element. The ions
pass through two slits (p.211), S_1 and S_2, in electrodes. The electrodes are given a negative
potential (voltage) to accelerate (p.219) the ions.
The ions then enter a semi-circular chamber
(p.143) which has a strong magnetic field acting
at a right angle to the path of the ions. The ions
follow a circular path, due to the magnetic field;
the lower the mass number of the ion, the smaller
is the radius of the path. The ions hit a
photographic plate and produce a mark. The
width of the mark corresponds to the abundance
(p.231) of an isotope. The spectrograph is cali-
brated (p.26) using carbon-12, so mass numbers
(p.113) can be read directly. The diagram shows
a photograph of the isotopes of magnesium. The
apparatus operates (p.157) with all the air
pumped out, forming a high vacuum. **mass
spectrography** (*n*).
mass spectrometer a piece of apparatus similar
to a mass spectrograph (↑); it measures
accurately the relative proportion (p.76) of
each isotope in a naturally-occurring element.

heat of reaction the heat energy given out, or taken in, when a chemical reaction takes place between the masses of reactants (p.62) shown by the equation (p.78) for the reaction. The heat energy is measured in joules (p.153).

heat of combustion the heat energy given out when one mole of a substance is completely burned in oxygen, e.g. ethyne (acetylene) is burned completely to carbon dioxide and water, 1 mole of ethyne produces 1558 kJ of heat energy. This is written as:
$C_2H_2 + \frac{5}{2}O_2 = 2CO_2 + H_2O$, $\Delta H = -1558$ kJ
The symbol (p.77) ΔH indicates (p.38) the heat change during the reaction. The negative sign shows heat is given out, i.e. lost by the reactants.

heat of neutralization the heat energy given out when one mole of hydrogen ions in an acid solution is neutralized by a base, with the reaction taking place in a dilute aqueous solution. In reactions between strong acids and strong bases, both reactants are completely ionized (p.123) and the reaction is between one mole of hydrogen ions and one mole of hydroxyl (p.132) ions. The heat of neutralization for strong acids with strong bases is 57.3 kJ; for weak acids or weak bases, the heat energy given out is less than 57.3 kJ, as energy is used to complete ionization of the acid or base.

heat of combustion of ethyne

heats of solution and dilution

heat of dilution the change in heat energy when a solution is diluted (p.81). *See diagram.*

heat of solution the heat energy given out, or taken in, when one mole of a substance is dissolved in such a large volume of water that further dilution (p.81) produces no heat change.

heat of formation the heat change when one mole of a compound is formed from its elements, under stated conditions of temperature and pressure.

determination of the heat of formation of methane (CH₄)

heat of ionization

heat of ionization the heat change necessary to produce complete ionization (p.123) of a ⁻substance in an aqueous (p.88) solution.

thermochemical equation an equation showing the chemical reaction between whole numbers of moles of the reactants, giving information about the states of the reactants, the temperature of measurement, and the heat change, e.g.

$$CH_{4(g)} + 2O_{2(g)} = CO_{2(g)} + 2H_2O_{(l)}$$
$$\Delta H_{298} = -889\,kJ$$

The reactants are gaseous methane and oxygen; the products are gaseous carbon dioxide and liquid water. The reaction between 1 mole of methane and two moles of oxygen gives out 889 kJ of heat energy at 298 K (25°C). For crystalline solids, the symbol (c) (p.77) is written, e.g. $S_{(c)}$ for crystalline sulphur.

stirrer

air-tight lid

platinum crucible

thermometer

inlet for oxygen

resistance wire to ignite substance

oxygen at 20 atmospheres pressure

steel calorimeter

known mass of water

bomb calorimeter

bomb calorimeter a device for measuring heats of combustion (p.146). It consists (p.55) of a thick-walled steel calorimeter, with an air-tight screw top. A known amount of substance is put in a platinum crucible and oxygen is pumped in to 20 atmospheres pressure. The substance is ignited (p.32) by a heated resistance wire. The heat evolved (p.40) heats the calorimeter, and thus the water. The temperature rise of the water is measured by a sensitive thermometer and the heat evolved is calculated, and hence the heat of combustion of the substance is found.

exothermic (*adj*) describes a reaction in which heat energy is given out.

endothermic (*adj*) describes a reaction in which heat energy is taken in.

Hess's law the heat energy given out, or taken in, in a chemical change is the same, no matter how the change takes place. For example, ammonium chloride solution can be made in two ways both starting with gaseous ammonia and gaseous hydrogen chloride:

(1) $NH_{3(g)} + aq \rightarrow NH_{3(aq)}$; $HCl_{(g)} + aq \rightarrow HCl_{(aq)}$
    $NH_{3(aq)} + HCl_{(aq)} \rightarrow NH_4Cl_{(aq)}$
    where (aq) is an aqueous solution.

(2) $HCl_{(g)} + NH_{3(g)} \rightarrow NH_4Cl_{(c)}$
    $NH_4Cl_{(c)} + aq \rightarrow NH_4Cl_{(aq)}$

The changes of heat energy are shown in the diagram opposite. Both ways of preparation produce the same result.

thermochemistry (*n*) the study of the changes in heat energy that take place during a chemical reaction. Heat energy given out heats the products, while heat energy taken in cools the products.

ΔH (kJ) $NH_{3(g)} + HCl_{(g)}$

−35.2

$NH_{3(aq)}$ + $HCl_{(g)}$

−176.1

−72.4

$NH_{3(aq)}$ + $HCl_{(aq)}$

$NH_4Cl_{(g)}$

−52.3   + 16.3

$NH_4Cl_{(aq)}$

rate of reaction the rate of reaction is measured by the rate at which the reactants are used up or by the rate at which the products are formed.

rate of reaction from formation of oxygen

rate constant if an amount x of a substance is used up in a reaction and the initial (p.85) amount was a, then the rate of reaction $(\frac{dx}{dt})$ at a given time is:$\frac{dx}{dt} \propto (a - x)$
i.e. the rate is proportional to $(a - x)$.
This can be written as: $\frac{dx}{dt} = k(a - x)$
where k is the *rate constant*.
The rate constant is independent of pressure in reactions in solution, dependent on pressure in gaseous reactions, but is changed by catalysis and temperature changes.

velocity constant = rate constant (↑).

law of mass action the rate of a reaction, at a constant temperature, is proportional to the product of the active masses of the reacting substances. In a reaction: $A + B \rightarrow$ products (A and B are two substances) the rate of reaction $\propto [A] \times [B]$, where $[A]$ represents the active mass of A. For gases, the active mass is measured by the partial pressure (p.108) of the gas; for solutions, it is measured by the concentration in moles per dm^3. With solids, the active mass is difficult to measure.

equilibrium (*n*) in a reversible reaction (p.64), the products start to react the moment they are formed. The rate of reaction of the reactants (p.62) is high as their concentration is high, while the rate of reaction of the products (p.62) is low because their concentration is low; this is because the law of mass action (p.149) is obeyed (p.107). Eventually, the two rates of reaction will be the same, provided the products are not removed (p.215), and the reactants and products will both be present. This is a state of *equilibrium* and thereafter there is no change in concentration of reactants and products.

dynamic equilibrium a state of equilibrium (↑) in which the rate of change of two processes (p.157) is equal and opposite, e.g. (a) in a closed vessel (p.25) the rate of evaporation (p.11) of a liquid, and the rate of condensation (p.11) of its vapour (p.11), are equal at a constant temperature; the vapour exerts (p.106) its vapour pressure (p.103) for that temperature. (b) phosphorus pentachloride (PCl_5) decomposes (p.65) on heating to phosphorus trichloride (PCl_3) and chlorine (Cl_2). In a closed vessel, PCl_3 and Cl_2 combine to form PCl_5. At equilibrium (↑) all three substances are present, with the two opposite chemical processes being in dynamic equilibrium.

$$PCl_5 \rightleftharpoons PCl_3 + Cl_2$$

equilibrium mixture the relative (p.232) concentrations of reactants and products when equilibrium (↑) is reached in a reversible reaction (p.64).

equilibrium constant in a reversible reaction (p.64):

$$pA + qB \rightleftharpoons rC + sD$$

where A, B, C, D are substances and p, q, r, s are mole fractions, the equilibrium constant is given by:

$$K = \frac{[C]^r \, [D]^s}{[A]^p \, [B]^q}$$

[A], [B], [C], [D] represent the concentrations of the substances. The value of *K* changes with temperature.

ethanol vapour

evaporation

ethanol liquid

condensation

dynamic equilibrium

rate of evaporation
= rate of condensation

reaction: **equilibrium**
ethanol + ethanoic acid
$\rightleftharpoons$ ethyl ethanoate + water

reactants and products condense (no substances lost from reaction)

ethanol
+
ethanoic acid

heat

equilibrium

ethanol
ethanoic acid
ethyl ethanoate
water

equilibrium mixture

Le Chatelier's principle whenever changes take place in the external conditions of a chemical reaction *in equilibrium*, then changes occur (p.63), if possible, in the reaction which tend (p.216) to counteract (p.216) the effect of those external changes. The principle can be illustrated from the reversible reaction:

$$N_{2(g)} + 3H_{2(g)} \rightleftharpoons 2NH_{3(g)} \quad \Delta H = -50\,kJ$$
1 vol.   3 vols   2 vols.

The reaction between nitrogen and hydrogen is exothermic (p.148) with 50kJ of heat energy given out. The decomposition of ammonia is endothermic (p.148) with 50 kJ heat absorbed. One volume of nitrogen reacts with 3 volumes of hydrogen to produce 2 volumes of ammonia, hence there are 4 volumes of the reactants and 2 volumes of the products. If an equilibrium mixture has its temperature raised the mixture absorbs heat to counteract the rise in temperature; this favours (p.214) the endothermic reaction, i.e. the decomposition of ammonia. Raising of the temperature quickens both rates of reaction, so equilibrium is reached more quickly. Increasing the pressure favours a smaller volume of gases, so the equilibrium moves towards forming more ammonia. The effect of external conditions on equilibrium is summarized in the table below.

REACTION:	A + B (larger volume)	C + D, (smaller volume)	+ heat given out
EXTERNAL CHANGE	RATE OF REACTION	EQUILIBRIUM MIXTURE	EQUILIBRIUM CONSTANT
increase of temperature	increased	more A + B	changed
decrease of temperature	decreased	more C + D	changed
increase of pressure	increased for gaseous reactions	more C + D	unchanged
decrease of pressure	decrease for gaseous reactions	more A + B	unchanged
addition of catalyst	increased	unchanged	unchanged

energy level an electron possesses energy according to its distance from the nucleus (p.110) of an atom; the nearer it is to the nucleus, the lower is the energy it possesses. This energy is the energy level of an electron.

ground state the condition of an atom (p.110) when all its extranuclear (p.113) electrons are in their positions of lowest energy, i.e. at their lowest energy levels (↑).

excitation (*n*) the changing of an atom from its ground state (↑) by supplying energy in the form of heat, radiation, or bombardment (p.143) by sub-atomic (p.110) particles. This causes one or more electrons to move to orbitals further from the nucleus. Such electrons may then enter into covalent bonds (p.136), or with sufficient energy, be removed from the atom, forming an ion (p.123). **excitatory** (*adj*).

energy barrier a measure of the quantity of energy that must be supplied in order that a chemical reaction can take place. If a lesser quantity of energy is supplied, there is no reaction.

activation energy the additional energy that must be supplied to reactants before a reaction can take place. The atoms are in a state of excitation (↑). The activation energy is the energy barrier to a reaction; until the activation energy is supplied, no reaction can take place.

activated state the state of molecules, atoms, or ions in which they are able to react.

reaction profile a diagram which shows the relation between the ground states (↑) of reactions and products, and the energy barrier to a reaction.

joule (*n*) the unit of energy and work. It is the work done when the point of application of a force of one newton is displaced through a distance of one metre in the direction of the force. It is also the work done when a current of 1 ampere flows between a voltage difference of 1 volt for 1 second. (A force of one newton gives a mass of 1 kg an acceleration of 1 ms^{-2}.) The symbol for joule is J.

calorie (*n*) a measurement of heat energy; it is that quantity of heat that will raise the temperature of 1 g of water through 1°C. The unit is no longer used. 1 calorie ≈ 4.18 joules.

electron-volt a unit of energy that is not included in S.I. units, but is useful in many chemical calculations on radiation (p.138) energy. It is the quantity of energy gained by an electron when it falls through a voltage difference of 1 volt.

$$1 \text{ electron-volt} = 1 \text{ eV}$$
$$= 1.6 \times 10^{-19} \text{ coulomb} \times 1 \text{ volt}$$
$$= 1.6 \times 10^{-19} \text{ joule.}$$

raw materials the materials which are used as the
starting point of chemical processes (p.157).
They include such materials as ores (↓), minerals
(↓), limestone, common salt, coal, mineral oil.

mineral (*n*) any material which occurs (↓)
naturally but does not come from animals or
plants. Examples of minerals are: metal ores
(↓), limestone (↓), rock salt (↓), coal (p.156),
mineral oil (p.156). A mineral has a known
chemical constitution (p.82) and definite
physical and chemical properties.

ore (*n*) a mineral (↑) which is a compound (p.8)
of a metal and from which the metal can be
extracted (p.164).

deposit (*n*) a mass of a mineral found in the earth,
in a large enough quantity to be worth mining (↓).

seam (*n*) a thick, fairly level, layer of a deposit
(↑) of a mineral (↑), e.g. a seam of coal.

lode (*n*) a large, thin deposit (↑) of an ore (↑)
which is between the walls of a deep crack in
the Earth's structure.

mine (*v*) to dig a hole in the ground to reach a
deposit, seam or lode (↑) and to get minerals
(↑) out. **mine** (*n*), **miner** (*n*), **mining** (*n*).

supply (*n*) the raw materials (↑), electric power,
fuel (p.160), necessary for the working of a
chemical process (p.157). **supply** (*v*).

reserve (*n*) a quantity of a material or mineral
which is not needed at present, but can be
used in the future, e.g. a reserve deposit of
mineral oil which can be used when present
supplies (↑) are finished.

occur[2] (*v*) to be in a particular place, e.g.
petroleum (p.160) occurs in Saudi Arabia in
great quantities. To contrast *occur* and *exist*
(p.213): sulphur *occurs* in large deposits in
North America; sulphur *exists* in two crystalline
(p.15) allotropes (p.118).

assay (*v.t.*) to make chemical tests (p.42) to find the mineral content (p.85) of a material; in particular to determine (p.222) the relative amount of a metal in an ore (↑).

native (*adj*) describes an element found in the Earth and not combined in a mineral, e.g. gold, copper, and sulphur are found uncombined in deposits (↑); they occur (↑) native, and are native elements.

blende (*n*) a mineral which is a sulphide (p.51) of a metal, e.g. zinc blende which is zinc sulphide.

pyrites (*n*) a mineral which is a sulphide of a metal, e.g. iron pyrites is iron sulphide.

chalk

chalk cliffs

chalk (*n*) a soft white rock consisting (p.55) of calcium carbonate, formed from the shells of shellfish. The shells were deposited (↑) on the sea-bed over a long time and pressure formed the rock from the shells.

limestone (*n*) a rock consisting (p.55) mainly of calcium carbonate, but sometimes containing magnesium carbonate. It is harder than chalk and is used as a building material. On heating, it is converted (p.73) to lime (p.169).

limestone in building

limestone

marble (*n*) a hard, crystalline (p.15) rock, formed from limestone (↑) by heat and pressure within the Earth. It consists (p.55) of calcium carbonate. Marble is used as a building material.

dolomite (*n*) a white crystalline (p.15) mineral (↑) consisting of calcium and magnesium carbonates. The mineral is found in great abundance (p.231).

marble

common salt the substance sodium chloride; it occurs (↑) in sea water which has a salt content (p.85) of about 3%. Common salt is obtained by evaporating large quantities of sea water.

rock salt a mineral consisting of sodium chloride. Large deposits (↑) are found in many countries. It is the raw material (↑) for the manufacture (p.157) of chlorine and many sodium salts (p.46).

sulphur (*n*) a yellow solid element, occurring (↑) native (↑) in the earth, also found in many ores (↑) such as blendes (↑) and pyrites (↑). It is the raw material (↑) for the manufacture (p.157) of sulphuric acid, the most important industrial (p.157) chemical.

carboniferous (*adj*) describes any material
having carbon in it, or being a form of coal (↓).
Also describes any material producing carbon
when heated, or any plant life which has
formed coal (↓).

carbonaceous (*adj*) describes any mineral
(p.154) containing carbon, but not carbonates,
e.g. coal (↓) is a carbonaceous mineral.

coal (*n*) a black, fairly hard mineral which occurs
in large deposits (p.154) and seams (p.154).
Coal has been formed from plant life which
existed (p.213) many millions of years ago.
Chemical action, heat and pressure of the
earth formed the mineral. It consists (p.55)
mainly of carbon, but contains many other
useful chemical compounds (p.8).

mineral oil a dark liquid mixture of various
compounds (p.8) occurring (p.154) in deposits
in many parts of the world; it varies widely in
composition (p.82) depending on the place
from where it comes. It is thought that mineral
oil has been formed from decayed plants and
animals. Deposits of mineral oil usually occur
with natural gas.

bitumen (*n*) a naturally occuring black material
which contains hydrocarbons (p.172). It is a
solid or a very thick liquid. Bitumen is also obtained as a product of distillation from coal (↑).

charcoal (*n*) a form of carbon obtained by
heating wood, or other plant materials or
animal materials, when no air is present
(p.217). Charcoal is black in colour and
generally very porous (p.15). Charcoal absorbs
(p.35) gases and decolorizes coloured liquids
if the colour is from organic (p.55) materials.

coke (*n*) a solid material obtained by heating
coal with no air present (p.217). It is mainly
carbon.

liquid air air is cooled below its critical temperature (p.104) and then liquefied (p.11). Liquid air
is the source (p.138) of nitrogen and oxygen.

silica (*n*) a very hard white substance, silicon
dioxide. It occurs in many minerals, e.g.
quartz, sand. It is the source of silicon, an
element used in electronic circuits.

oil derrick

rock strata

impervious rock cap

gas under pressure

mineral oil

drill

porous sandstone

industrial process

sodium chloride

electrolysis of solution

hydrogen

chlorine

by-product

by-product

sodium hydroxide

main product

industrial process for
the manufacture of
sodium hydroxide

process (*n*) (1) in chemistry, a method of making a substance in large quantities; the method includes a description of the chemical reaction, the conditions (p.103) under which the reaction takes place and the plant (↓) needed for the process. (2) the different methods of preparing (p.43) and isolating (p.43) a substance, e.g. distillation, sublimation. (3) generally, a process is a number of events taking part in a connected and continuous change. **process** (*v.t.*).

plant (*n*) all the building, pipes, furnaces (p.164), machinery, special apparatus (p.23) and devices (p.23) used to make substances in large quantities for industry (↓).

industrial (*adj*) describes anything connected with the making of chemical substances, machinery, or materials in large quantities. Industrial chemicals are divided into heavy chemicals (used in other industrial processes), pharmaceutical chemicals (p.20) (used in curing diseases), fine chemicals (used in chemical analysis (p.82)), fertilizers and plastics. **industry** (*n*).

operate (*v.t.*) to make an industrial (↑) process work, e.g. to operate a process for the manufacture (↓) of ammonia is to provide the reactants and obtain the products using the necessary plant (↑). **operation** (*n*).

carry out to work on a process, an experiment or an investigation and to complete it.

manufacture (*v.t.*) to make chemical substances, materials, machinery by an industrial (↑) process (↑). **manufacture** (*n*).

main product the product for which a particular process (↑) is planned. In some processes it is the only product, e.g. sulphuric acid in the Contact process (p.166) is the only product.

by-product (*n*) a substance produced in the manufacture (↑) of a main product (↑); it has an industrial (↑) use, e.g. in the Castner-Kellner process (p.169) chlorine is a by-product.

waste product a substance that is produced in a chemical process and which has no industrial (↑) use, e.g. calcium chloride, produced in the Solvay process (p.169), has no industrial use.

end product a product formed after an original (p.220) substance has taken part in several reactions or processes, e.g. petrol is one end product from the refining (↓) of petroleum.

mill (*n*) (1) a machine for breaking large lumps (p.13) of solid materials and turning the material into a powder (p.13), e.g. a mill for making flour. (2) a work place where cloth is made, e.g. a cotton mill.

pulverize (*v.t.*) to turn pieces of solid into a powder (p.13) by hitting them repeatedly. A mill (↑) can pulverize a solid. **pulverization** (*n*).

spraying (*n*) the action of sending many small drops of liquid from small holes in a pipe so as to cover a large space. **spray** (*n*), **spray** (*v.t.*).

lixiviation (*n*) the process of removing one substance from a mixture of substances by using a suitable solvent (p.86), e.g. salts containing iodine are obtained from heated seaweed by lixiviation with water. **lixiviate** (*v*).

leaching (*n*) the removal (p.215) of soluble (p.17) substances by water from a mixture of solids when the water washes out the substance. To contrast *lixiviation* and *leaching*: if a mixture is treated (p.38) with water, the process is *lixiviation*; if water flows through the mixture to wash away a substance the process is *leaching*. Solvents other than water can be used in leaching. **leach** (*v*).

slaking (*n*) adding water to quicklime (calcium oxide) to form slaked lime (calcium hydroxide). **slake** (*v*).

scum (*n*) any solid substance, especially dirt or waste (p.170), which floats on the surface of a liquid, e.g. impurities floating on the surface of a molten (p.10) metal.

sludge (*n*) soft wet solid material, usually waste or unwanted material.

froth flotation a process for the separation (p.34) of ores (p.154) from earthy material. Oil and water form a froth (p.100) with particles of ore trapped in the liquid-air interfaces (p.18) of the bubbles (p.40). The foam is stabilized by frothing agents (p.63). The ore is removed with the froth.

spraying

spray of water

scum/sludge

scum floating and sludge settling in a liquid

scum

water (or other liquid)

sludge

refining copper by **refine**
electrolysis

dross

stage (*n*) a length of space or time in a process,
e.g. (a) the primary, intermediate and secondary
stages of an education system (p.212); (b) the
stages in the Bosch process (p.168) for
manufacturing hydrogen are (i) mixing water
gas and steam, (ii) passing the gases over a
catalyst, (iii) removing carbon dioxide.

refine (*v*) (1) to remove impurities from a substance, e.g. to refine copper by electrolysis
(p.122). (2) to separate the constituents (p.54)
of a mixture to get pure specimens of some, or
all, of the constituents, e.g. to refine petroleum
(p.160) to get petrol, kerosene, etc. (3) to make
a method, apparatus, or a technique more
efficient or more suitable for its purpose, e.g.
(a) a refined method of determining the equilibrium constant of a reaction; (b) a more refined
apparatus for carrying out (p.157)
chromatography. **refinement** (*n*).

liquation (*n*) a method of refining (↑) a metal in
which a mixture of metals is heated until one
metal melts (p.10) and flows away from the
mixture.

dross (*n*) the impurities, or other waste material,
that floats on top of molten (p.10) metal as a
scum (↑). Dross is removed (p.215) in refining
(↑) metals.

optimum (*adj*) describes a condition which is
the most favourable (p.214) for a reaction, e.g.
the optimum temperature for the Contact
process (p.166) is 500°C. At this temperature
the best yield (↓) is obtained.

yield (*n*) the quantity of a product obtained from
an industrial (p.157) process or preparation
(p.43). The actual yield is often compared, as
a percentage, with the theoretical
stoichiometrical (p.82) yield, e.g. the yield in
the Haber process is only about 6%, i.e. only
6% of nitrogen and hydrogen combine to form
ammonia. **yield** (*v*).

replenish (*v*) if a material is being used in a
process, further quantities have to be added to
replace (p.68) the losses. To replenish is to
make these additions of the material.
replenishment (*n*).

fuel (*n*) any material or substance that is burned to give heat, and through heat, to give power. Common fuels in industry are: coal, coke, kerosene, fuel oil, coal gas, natural gas.

petroleum (*n*) mineral oil (p.156) consisting (p.55) of hydrocarbons (p.172) and some compounds of sulphur and nitrogen. Petroleum is fractionally distilled (p.201) to obtain fractions (p.202), each of which has a particular use.

petrol (*n*) a volatile (p.18) liquid of low boiling point (p.12) obtained from petroleum (↑) by distillation (p.33). Petrol distils between 20°C and 150°C. It is a mixture of hydrocarbons (p.172) from hexane to decane. Petrol is highly flammable (p.21); it is used in internal combustion engines.

gasoline (*n*) name for petrol in the U.S.A.

kerosene (*n*) a volatile liquid similar to petrol (↑) but with a higher boiling point. It is obtained from petroleum (↑) by fractional distillation; kerosene distils between 150°C and 250°C. It is used for heating, lighting and in jet engines.

fractional distillation of petroleum

petrol	fuel in motor car engines
kerosene	for heating and lighting fuel in jet engines
gas oil	fuel in diesel engines
lubricants	lubricating moving parts of machinery
pitch	surfacing roads sealing roofs

asphalt (*n*) a black, sticky, solid material consisting (p.55) mainly of hydrocarbons (p.172). It occurs naturally in various places and is also found in some kinds of petroleum (↑). Asphalt is used for road surfaces and for making roofs waterproof.

naphtha (*n*) an organic (p.55) liquid obtained by dry distillation (p.203) of coal or wood. It can contain (p.55) a number of different compounds, but has no particular composition (p.82). Naphtha is a flammable (p.21), volatile (p.18) liquid.

paraffin[1] (*n*) another name for kerosene (↑); paraffin oil is also used as a name.

gas oil a liquid obtained from petroleum (↑) by fractional distillation. It is less volatile than kerosene and distils between 250°C and 300°C. Gas oil is used in diesel engines.

lubricants (*n.pl.*) various kinds of lubricating oils. Lubricants are used to reduce friction. They are distilled from petroleum at between 300°C and 400°C. Lubricants with a high boiling point are soft solids such as petroleum jelly. Lubricants with lower boiling points are thick, sticky liquids.

doctor solution a solution which is used to remove bad-smelling compounds from petroleum (↑).

pitch (*n*) a black, sticky liquid left after fractional distillation of petroleum (↑). It is the same material as asphalt (↑).

tar (*n*) a thick, black, sticky liquid formed during the destructive distillation (p.203) of coal. It contains many different organic (p.55) substances from which many useful compounds are manufactured (p.157).

scrubber (*n*) a device for removing (p.215) ammonia and benzene from impure coal gas. The impure gas is passed up the scrubber, and a spray of water flows down the scrubber, washing out the ammonia and benzene. Ammoniacal liquor, containing these two substances, collects in the scrubber. **scrub** (*v*).

gasometer (*n*) a very large iron vessel (p.25) for the storing of coal gas.

vat (*n*) a large, open vessel (p.25) used for making wine or soap, dyeing cloth, etc.

dye (*n*) a substance used to give colour to cloth, plastics or paper. Some dyes are made from plants, but most are synthetic (p.200). **dye** (*v*).

pigment (*n*) a solid substance used to give colour to paint or varnish. A pigment is not soluble in water. Pigments can be either organic (p.55) or inorganic (p.55) substances.

mordant (*n*) a substance which is used with dyes (↑) that do not dye a material directly. The material is treated (p.38) first with the mordant; the treated material is then put in a vat (↑) with a dye. With acid dyes, the mordant is aluminium or tin hydroxide. With basic dyes, vinegar (ethanoic acid) or tannic acid is used as a mordant. The mordant causes the dye to dye the material.

soap (*n*) any salt of a metal and a fatty acid. The water-soluble soaps are the sodium and potassium salts; sodium forming hard soaps and potassium forming soft soaps. These soaps have a cleansing action, removing dirt from surfaces. Other soaps have different properties. The most common fatty acids in water-soluble soaps are stearic acid ($C_{17}H_{35}COOH$) and palmitic acid ($C_{15}H_{31}COOH$). **soapy** (*adj*).

salting out the addition of a concentrated solution of sodium chloride to a solution of an organic (p.55) compound in water or ethanol, in order to throw the organic compound out of solution.

impure coal gas

scrubbed coal gas

water

water spray (removes ammonia and benzene)

ammoniacal liquor

scrubber

vat

gasometer iron vessel

30 m

coal gas

water seal

blast furnace a vertical furnace (p.164) for obtaining iron from its ores. A strong current (a blast) of air is blown up the furnace. Coke (p.156) is mixed with limestone (p.155) and the iron ore. The furnace is kept at a temperature of about 1800°C in the hottest part. The coke is converted to carbon monoxide in two stages:
$$C + O_2 \rightarrow CO_2 \qquad CO_2 + C \rightarrow 2CO$$
The carbon monoxide reduces the iron ore to iron. Molten (p.10) iron runs out of the bottom of the furnace.

tuyere (*n*) a pipe, usually made of copper, through which air is blown into a blast furnace.

pig iron (*n*) iron, straight from a blast furnace (↑), which has cooled in moulds (p.210). It contains (p.55) 2 – 5% carbon and smaller quantities of other impurities making it hard and brittle (p.14).

cast iron pig iron which has been melted, mixed with steel scrap (p.171), and cooled in moulds (p.210) to give it a shape. Cast iron is impure iron and is hard and brittle.

wrought iron wrought iron is obtained from pig iron by heating it in a reverberatory furnace (p.164) with limestone and stirring the molten (p.10) mass with long iron rods. Wrought iron is almost pure, containing (p.55) less than 0.2% carbon. It is soft, malleable and easily welded.

puddling process cast iron is fused in a reverberatory furnace (p.164) with haematite (an ore of iron oxide) lining the hearth (p.164). The oxygen in haematite oxidises the carbon in cast iron and forms nearly pure wrought iron (↑).

slag (*n*) a waste product (p.157) from a blast furnace (↑). It consists (p.55) of calcium silicate formed by the reaction between limestone and the earthy parts of iron ore.

clinker (*n*) a hard mass of material which is not combustible (p.58), formed in furnaces and boiler fires. It consists of silicates formed by fusion (p.32) of earthy materials in fuels and ores.

roast (*v*) to heat metals or ores in air at a temperature too low for fusion. The roasting removes impurities (p.20) by oxidation (p.70) with atmospheric oxygen. Roasting removes (p.215) sulphur and sulphur dioxide from sulphide ores.

iron ore
coke
limestone

iron ore
reduced

hot
air

tuyere

molten
slag

molten iron

slag

pig-iron
out

blast furnace

smelt (v) to separate a metal from its ore by heating the ore with a suitable reducing agent (p.71). The metal becomes molten (p.10) and impurities separate from the metal, e.g. iron ore is smelted in a blast furnace (p.163). **smelting** (n).

extract (v) to obtain an element from the Earth, either as a native element or by mining the ore and obtaining the element by chemical action, e.g. extracting aluminium by mining bauxite (aluminium oxide) and electrolyzing it to obtain aluminium. **extraction** (n).

reverberatory furnace

furnace (n) a brick construction inside which great heat is produced by burning fuels (p.160) or by electricity; mainly used in the extraction of metals.

flue (n) a pipe leading hot air and smoke away from a fire.

reverberatory furnace a furnace (↑) with a low roof above the hearth (↓) so that the heat from the fire is directed onto the reactants on the hearth. Reverberatory furnaces are used for smelting ores.

hearth (n) the floor on which a fire burns or a reaction takes place in a furnace.

open-hearth furnace a shallow, rectangular furnace (↑) which is heated by burning a gaseous (p.11) fuel. The hot gases from the burning fuel pass over the hearth (↑) on which an ore is placed. The hearth is covered with limestone or dolomite (p.155).

open-hearth furnace

ash (n) a residue (p.31) which is a powder, left after a material has been burned completely, e.g. the ash left after wood or plants have been burned. **ashen** (adj).

converter (n) a large iron vessel in which pig iron (p.163) is converted (p.73) to steel in the Bessemer process (↓).

metallurgy (n) the science of the extraction of metals from their ores (p.154), refining (p.159) the metals, and forming alloys (p.55). **metallurgist** (n).

weld (v) to join together two pieces of metal by (a) heating them so that they melt (p.10) and join or (b) by beating them with a hammer to make them soft enough to join together. **welding** (n).

steel (*n*) steel is obtained from pig iron by first oxidizing all the impurities away and then adding a known amount of carbon to the molten (p.10) iron. Other metals, such as manganese or chromium, can be added to make different kinds of steel. Steel contains between 0.15% to 1.5% carbon, depending on the kind of steel that is wanted. Steel is hard and elastic.

quench (*v*) to harden steel by heating it until red-hot, then quickly putting it into cold water or cold oil. **quenching** (*n*).

temper (*v*) quenched (↑) steel is heated to a temperature of 200°C-600°C, depending upon the kind of tempering needed. The steel is kept at that temperature for 30 minutes and then allowed to cool. Tempering makes a steel elastic as well as hard. **tempering** (*n*).

pickling (*n*) steel is put in vats with concentrated sulphuric acid. This removes rust (p.61) and any other surface impurities. The steel can then be galvanized (p.166), tinned (p.166), or painted. **pickle** (*v*).

Bessemer process a process to make steel (↑) from pig iron (p.163) or cast iron (p.163). A converter (↑) is lined with calcium and magnesium oxides; pig iron and scrap (p.171) steel are added as molten (p.10) metal. Air is blown through the metal oxidizing all the impurities. Carbon, as needed, is added to make the steel.

Siemens-Martin process pig iron (p.163), scrap (p.171) steel and haematite (iron (III) oxide) are heated in an open-hearth furnace (↑). The hearth (↑) is lined with limestone. The proportion of iron haematite and steel are calculated to give the correct proportion of carbon in the product.

open-hearth process another name for the Siemens-Martin process.

Linz-Donawitz process molten, impure pig iron (p.163) from the blast furnace is used to make steel. Oxygen is blown over the molten (p.10) iron causing the oxidation of carbon and other impurities. Lime powder is added to form a slag (p.163). This is a rapid process (it takes 10 – 20 minutes) and steel as pure as the steel from the open-hearth process is formed.

quenching steel

Bessemer converter

spelter (*n*) zinc that has not been refined
(p.159); it contains other metals (e.g. lead) and
some impurities.

galvanize (*v*) to cover an iron surface with a coat
(p.127) of zinc, usually spelter (↑); a metal object
is dipped into molten (p.10) zinc. **galvanized** (*adj*).

sherardize (*v*) to cover an iron surface with a coat
(p.127) of zinc. A metal object is heated in a
closed vessel containing zinc dust at a
temperature just below the melting point of zinc.

tin (*v*) to coat (p.127) an iron article with a thin
layer of tin. The iron is first pickled (p.165) and
then dipped in molten (p.10) tin. **tinning** (*n*).

contact process a process for making concentrated sulphuric acid by converting (p.73)
sulphur dioxide to sulphur trioxide and then
converting the trioxide to sulphuric acid. The
process is shown in the diagram below. Sulphur
dioxide and oxygen (from air) are passed over
a catalyst (p.72) of vanadium pentoxide at
500°C, forming sulphur trioxide. The sulphur
trioxide is absorbed in concentrated sulphuric
acid forming oleum (↓), which is diluted to
form sulphuric acid of 98% concentration.

contact process

steam 77% acid

$SO_2 + H_2O + NO_2$
$\updownarrow$
$H_2SO_4 + NO$
$2NO + O_2 \rightarrow 2NO_2$

lead chamber

$SO_2 + NO_2$

Glover tower

Gay-Lussac tower

NO_2

nitrogen dioxide dissolved in sulphuric acid

gases cooled acid concentrated

acid diluted from 77% to 65%

$H_2SO_4 + NO_2$

SO_2

65% $H_2SO_4 + NO_2$

sulphur

77% H_2SO_4

lead chamber process

burner

lead-chamber process an older process for the manufacture of concentrated sulphuric acid. Sulphur dioxide is oxidized to sulphur trioxide by nitrogen dioxide: $SO_2 + NO_2 \rightarrow SO_3 + NO$ The nitrogen oxide (NO) is oxidized by oxygen to nitrogen dioxide; the nitrogen compounds act like a catalyst. The process is shown in the diagram. This process produces a 77% concentration of acid, not as concentrated and not as pure as the acid from the Contact process, but still useful for manufacturing processes.

oleum (n) concentrated sulphuric acid containing dissolved sulphur trioxide; also known as fuming sulphuric acid.

spent oxide iron (III) oxide is used to remove sulphur compounds from coal gas; the oxide eventually contains a high percentage of sulphur and is no longer of use for purifying (p.43) the coal gas; it then becomes spent oxide, i.e. it has been used up. Spent oxide is burned to form sulphur dioxide for use in the manufacture of sulphuric acid.

water gas a mixture of hydrogen and carbon
monoxide formed by blowing steam over red-
hot coke (p.156). The reaction cools the coke.

producer gas a mixture of nitrogen and carbon
monoxide formed by blowing air through coke
(p.156). The reaction heats the coke.

semi-water gas a mixture of hydrogen, carbon
monoxide and nitrogen, formed by blowing
steam and air alternately over coke (p.156). One
reaction heats, and the other reaction cools,
the coke, so its temperature is maintained.

Bosch process a process for making hydrogen
from water gas (↑). Water gas and steam are
passed over a catalyst of iron with traces
(p.20) of chromium (III) oxide as a promoter
(p.72). At 450°C, the reaction is:

$$H_2 + CO + H_2O \rightleftharpoons CO_2 + 2H_2$$

The carbon dioxide is removed by washing
with hot potassium carbonate solution. Natural
gas, containing methane, is now used in this
process instead of water gas.

mercury cathode cell a method manufacturing
sodium hydroxide and chlorine from sodium
chloride. A solution of sodium chloride is electrolyzed (p.123) using carbon anodes (p.123)
and a mercury cathode. Chlorine is taken
away by pipes from the cell. The cathode
forms a sodium amalgam (p.55), which is removed and taken to another tank where sodium
hydroxide is formed and hydrogen set free.

mercury cathode cell

carbon anodes

chlorine

brine

spent brine

pump

mercury cathode

sodium amalgam

sodium reacts with water
$2Na + H_2O$
$\rightarrow 2NaOH + H_2$

hydrogen

sodium hydroxide solution

water

sodium amalgam

mercury

lime (*n*) lime is either **quicklime**, calcium oxide, or **slaked lime**, calcium hydroxide.

soda (*n*) soda is either **caustic soda**, sodium hydroxide, or **washing soda**, sodium carbonate.

brine (*n*) a solution of sodium chloride.

Castner-Kellner process another name for the mercury cathode cell (↑).

Kellner-Solvay process another name for the mercury cathode cell (↑).

Solvay process a process for the manufacture of sodium carbonate ($Na_2CO_3.10H_2O$) using sodium chloride and calcium carbonate as raw materials. Brine (↑) is saturated with ammonia and passed down a column, up which is passed carbon dioxide gas. The reaction which takes place is:
$$NaCl + NH_3 + H_2O + CO_2 \rightarrow NaHCO_3 + NH_4Cl$$
The sodium hydrogen carbonate ($NaHCO_3$) is heated to form sodium carbonate and carbon dioxide. The sodium carbonate is recrystallized to form washing soda ($Na_2CO_3.10H_2O$). The carbon dioxide and ammonia from the ammonium chloride are put back into the cycle (p.64) of operations (p.157) as shown in the diagram below. Calcium chloride is produced as a waste product.

Solvay process

Haber process a process for the manufacture of
ammonia by synthesis (p.200) from nitrogen
and hydrogen. The two elements combine in
the presence of a catalyst (iron with potassium
oxide as a promoter (p.72)) under high pressure
at 450°C – 500°C. The reaction is reversible
(p.64) and only 6% of the elements combine.
The ammonia is liquefied by cooling and then
removed (p.215); the uncombined gases are
passed back to the catalyst chamber.

kiln (*n*) a furnace (p.164) for making bricks,
ceramics (↓) and heating chalk to form lime
and carbon dioxide.

spent (*adj*) describes something which has
been used and is finished; also something
which has had an important constituent used,
so it can no longer carry out its purpose, e.g.
spent brine which is too dilute to be of use.

waste (*adj*) describes something that has been
made but is not wanted, e.g. paper that is no
longer wanted is waste paper, the paper is
usable but not wanted for its original purpose

obsolete (*adj*) describes something which is usable but has been replaced (p.68) by something which is better, and so is no longer used, e.g. retorts (p.28) are obsolete, as distillation flasks (p.28) have replaced them.

scrap (*n*) describes something which is no longer used, e.g. when a car is no longer used it becomes scrap. The steel in machines which are no longer used, is scrap steel. Metals are most commonly described as scrap.

Thermit process a process for the extraction (p.34) of chromium, manganese or tungsten from their oxides or for welding (p.164) iron objects. Aluminium powder is mixed with the metal oxide and a piece of magnesium is set alight to start the reaction. The aluminium reduces the oxide and, in the case of iron oxide, produces molten (p.10) iron to weld two iron objects together.

detergent (*n*) any substance used for removing (p.215) dirt. Soaps are detergents. Other substances used as detergents are made from sulphonic acids. Their molecules consist of hydrocarbon chains attached to acidic groups. *See sulphonate (p.193).*

vulcanization (*n*) the process for changing rubber from a weak material into a hard, strong material, e.g. the rubber used in car tyres has been hardened by vulcanization. Vulcanization is usually carried out by heating rubber with sulphur. **vulcanize** (*v*).

ceramics (*n*) the manufacture of earthenware and porcelain objects. **ceramic** (*adj*).

tamp (*v*) to push soft, powdery solids into a hole and to fill the hole completely, e.g. in the Thermit process, the aluminium powder and metal oxide is tamped into the crucible. **tamping** (*n*).

heavy chemical a chemical in great demand for industrial (p.157) processes and thus manufactured (p.157) in large quantities. Such chemicals are often not very pure (p.20). Examples of heavy chemicals include sulphuric acid, nitric acid, lime and sodium carbonate. *See fine chemicals (p.20).*

mixture of aluminium powder and metal oxide

powder tamped down

piece of magnesium (to start the reaction)

crucible

plug to release molten metal

Thermit process

ceramic objects

hydrocarbon (*n*) an organic (p.55) compound containing (p.55) only the elements carbon and hydrogen; all hydrocarbons are covalent (p.136) compounds. Hydrocarbons include alkanes (↓), alkenes (↓), alkynes (p.174) and benzene (p.179) compounds.

series[2] (*n*) a group of organic (p.55) compounds which all have similar chemical properties; physical properties showing a regular change with an increasing number of carbon atoms; can be prepared (p.43) by similar chemical methods; described by a general formula. The compounds are homologous (↓).

homologous (*adj*) describes structures which are alike without being exactly the same. In homologous series (↑), all the compounds have a general formula (p.181), the same functional group (p.185) and a gradation of properties (p.9), e.g. the alkanes (↓) form an homologous series; the carboxylic acids form an homologous series, each acid having the functional group − COOH. **homologue** (*n*).

alkane (*n*) a hydrocarbon (↑) with the general formula (p.181) $C_n H_{2n + 2}$. The alkanes form an homologous (↑) series; the first four members are: CH_4 (n = 1); C_2H_6; C_3H_8; C_4H_{10} (n = 4). Alkanes are either straight chain (p.182) or branched chain compounds. The first four members of the alkanes are methane, ethane, propane and butane; thereafter they have a Greek number prefix for the number of carbon atoms, followed by -*ane*, e.g. hexane, C_6H_{14}. The alkanes have properties similar to methane (↓), becoming less reactive (p.62) with an increasing number of carbon atoms, and passing from gases through liquids to solids for members with a large number of carbon atoms. They are saturated (p.185) organic compounds.

methane (*n*) an alkane (↑) of formula CH_4. It is an odourless (p.15), colourless (p.15), flammable (p.21) gas. The hydrogen atom can be replaced by halogens (p.117), otherwise methane is unreactive (p.62).

homologous series of alkanes

formula	melting point °C	boiling point °C
CH_4	−183	−162
C_2H_6	−184	−89
C_3H_8	−188	−42
C_4H_{10}	−138	−1
C_5H_{12} ·	−130	36
C_6H_{14}	−95	69
C_7H_{16}	−91	98
C_8H_{18}	−57	126

carbon atom

hydrogen atom

formula

$$H - C - H$$

(with H above and H below)

methane

ethane

formula

H H
| |
H — C — C — H
| |
H H

● hydrogen atom
● carbon atom

ethane (*n*) the second member of the alkanes
(↓), formula C_2H_6. Its properties are similar to
those of methane (↑), but it is less reactive,
has a higher boiling point and a greater density.
paraffin[2] (*n*) traditional name (p.44) for an alkane
(↑).

propene

double bond

H H
| |
H — C — C = C
| | |
H H H

formula

H H H H
| | | |
H — C — C — C = C
| | |
H H H

$CH_3 CH_2 CH = CH_2$
a butene
but–1–ene

formula

formula

but–2–ene
a butene
$CH_3 CH = CH CH_3$

H H H H
| | | |
H — C — C = C — C — H
| |
H H

alkene (*n*) a hydrocarbon (↑) with the general
formula (p.181) C_nH_{2n}. The alkenes form an
homologous (↑) series; the first three members
are: C_2H_4 (n = 2); C_3H_6; C_4H_8 (n = 4). Alkenes
are either straight chain (p.182) or branched
chain compounds. The first four members of
the alkenes are: ethene, propene, butene and
pentene; thereafter they have a Greek number
prefix for the number of carbon atoms, followed
by -ene, e.g. hexene, C_6H_{12}. The alkenes have
properties similar to ethene (p.174), becoming
less reactive with an increasing number of
carbon atoms, and passing from gases through
liquids to solids for members with a large
number of carbon atoms. They are unsaturated
(p.185) organic compounds.
olefine or **olefin** (*n*) the traditional name (p.44)
for an alkene (↑).

ethene (*n*) an alkene of formula C_2H_4. It is a colourless (p.15) gas with a sweet smell. The two carbon atoms are joined by a double bond (p.181) which makes ethene reactive (p.62). It undergoes (p.213) addition reactions (p.188), polymerization (p.207) and combustion (p.58).

ethylene (*n*) traditional name (p.44) for ethene (↑).

alkyne (*n*) a hydrocarbon (p.172) with the general formula (p.181) C_nH_{2n-2}. The alkynes form an homologous (p.172) series; the first three members are: C_2H_2 (n = 2); C_3H_4; C_4H_6 (n = 4). Alkynes are either straight chain (p.182) or branched chain compounds. The first three members of the alkynes are ethyne, propyne and butyne; thereafter they have a Greek number prefix for the number of carbon atoms, followed by -*yne*, e.g. hexyne, C_6H_{10}. The alkynes have properties similar to ethyne (↓), becoming less reactive with an increasing number of carbon atoms. They are unsaturated (p.185) organic compounds.

acetylenes (*n.pl.*) traditional name (p.44) for the alkynes (↑).

double bond

formula

ethene

ethyne

triple bond

formula   H — C ☰ C — H

ethyne (*n*) an alkyne (↑) of formula C_2H_2. It is a colourless (p.15) gas with a sweet smell. Ethyne burns with a bright, white flame (p.58) and is used for lighting. The two carbon atoms are joined by a triple bond (p.181) which makes ethyne very reactive (p.62). Ethyne undergoes (p.213) addition reactions (p.188) and polymerization (p.207). With alkali metals (p.117), acetylides (p.49) are formed.

acetylene (*n*) traditional name (p.44) for ethyne.

propan-2-ol

CH_3 formula
$CHOH$ a secondary alcohol
CH_3

2 methylpropan-2-ol

CH_3 formula
$CH_3 - C - OH$ a tertiary alcohol
CH_3

● carbon atom
● hydrogen atom
○ oxygen atom

ethanal

an aldehyde

formula $CH_3 CHO$

alcohol[1] (*n*) an organic (p.55) compound which contains one or more hydroxyl (p.185) groups. Alcohols can be formed from alkanes (p.172) by substituting (p.188) a hydroxyl group for a hydrogen atom; they form homologous series (p.172) corresponding (p.233) to the alkanes. Alcohols are classified as primary, secondary or tertiary alcohols depending on the position of the hydroxyl group. For primary alcohols, the group is $- CH_2OH$; for secondary alcohols it is $> CHOH$; for tertiary alcohols it is $\gg COH$. Alcohols are also classified according to the number of hydroxyl groups in the molecule:

monohydric alcohol	CH_3CH_2OH	ethanol
dihydric alcohol	CH_2OH $\|$ CH_2OH	ethane-1,2-diol (glycol)
trihydric alcohol	CH_2OH $\|$ $CHOH$ $\|$ CH_2OH	propane-1,2,3-triol (glycerol)

Alcohols react with alkali metals (p.117) evolving (p.40) hydrogen and forming an alkoxide; they burn readily and are oxidized to aldehydes (↓), ketones (p.176) or carboxylic acids (p.176) depending on the alcohol and the strength of the oxidizing agent. With organic acids alcohols form esters (p.177). The hydroxyl group can be replaced by a halogen (p.117).

ethanol a primary alcohol

$H - \underset{\underset{H}{|}}{\overset{\overset{H}{|}}{C}} - \underset{\underset{H}{|}}{\overset{\overset{H}{|}}{C}} - OH$ formula

alcohol[2] (*n*) trivial name (p.44) for ethanol.
aldehyde (*n*) an organic (p.55) compound which contains the functional group $- CHO$. An aldehyde is formed as a first oxidation product (p.62) from the corresponding (p.233) alcohol (↑), e.g. ethanol is oxidized to ethanal. The name of an aldehyde is derived (p.106) from the corresponding alcohol by changing -*ol* to -*al*. Aldehydes are reduced to alcohols; they form addition (p.188) compounds with sodium hydrogen sulphite, hydrogen cyanide and other compounds.

ketone (*n*) an organic (p.55) compound containing the functional group > CO Ketones are pre-pared (p.43) from the corresponding secondary alcohol (p.175) by oxidation. They are good solvents for organic compounds and are less reactive than aldehydes (p.175), although their reactions are similar to those of aldehydes.

carboxylic acid an organic (p.55) compound containing the functional group − COOH. Carboxylic acids are prepared (p.43) from the corresponding alcohol (p.175) by complete oxidation; they are named from that alcohol, e.g. ethanol on oxidation forms ethanoic acid, the ending -*ol* is replaced by -*oic*. The

ketone

dimethyl ketone propanone

$$CH_3 - C - CH_3$$
formula $\overset{\|}{O}$

$$CH_3CH_2OH \xrightarrow{\textbf{(O)}} CH_3COOH$$
ethanoic acid

ethanoic acid

formula $CH_3 - C \overset{\ominus}{\underset{OH}{<}}$

carboxylic acid

carboxylic acids are all weak acids (p.45). With alcohols, they form esters (↓); the hydroxyl group (p.185) can be replaced by a halogen (p.117); dehydration forms an acid anhydride (↓). The carboxylic acids form an homologous series (p.172).

dicarboxylic acid an organic (p.55) compound with two functional groups of − COOH. The simplest member is ethane di-oic acid (traditional name (p.44) oxalic acid). The dicarboxylic acids are stronger acids than the carboxylic acids; they are dibasic (p.46).

acid anhydride an organic (p.55) compound prepared from the corresponding (p.233) acid by dehydration or by the action of an acyl chloride (↓) on the sodium salt of the acid.

acyl chloride an organic (p.55) compound containing the functional group − COCl; it is the functional group − COOH with the hydroxyl group replaced (p.68) by chlorine. Acyl chlorides are prepared from the corresponding carboxylic acid (↑) by the action of phosphorus pentachloride. They are very reactive (p.62) and react with water to give the corresponding carboxylic acid.

formula

COOH
|
COOH

ethanedioic acid
oxalic acid

dicarboxylic acid

acyl chloride

ethanoyl chloride
acetyl chloride

formula CH_3COCl

● carbon atom
● hydrogen atom
○ oxygen atom
○ chlorine atom

ester

formula

CH₃COOCH₃

methyl ethanoate

a fat

ether (dimethyl ether)

methoxymethane

formula

CH₃ — O — CH₃

alkyl halide

chloroethane

formula **CH₃CH₂Cl**

amide ethanamide

formula

CH₃CONH₂

ester (*n*) an organic (p.55) compound formed when an alcohol (p.175) reacts with a carboxylic acid (↑). The hydrogen of the acid is replaced (p.68) by an alkyl (p.180) or any aryl (p.180) group. The reaction is reversible (p.64), e.g.

$$CH_3CH_2OH + CH_3COOH \rightleftharpoons CH_3COOC_2H_5 + H_2O$$

ethanol + ethanoic acid ⇌ ethyl ethanoate + water

An ester can also be formed with an inorganic (p.55) acid. Esters have fragrant (p.22) odours and are the odours of flowers and fruit; they are not very reactive, but undergo (p.213) hydrolysis (p.66).

fat (*n*) an ester (↑) of propane-1,2,3-triol (glycerol) and different carboxylic acids (↑). In fats, the acids are saturated (p.185), straight chain (p.182) compounds. In oils, the acids are unsaturated (p.185), straight chain compounds. The important acids in animal and plant fats are palmitic acid ($C_{15}H_{31}COOH$), stearic acid ($C_{17}H_{35}COOH$) and lauric acid ($C_{11}H_{23}COOH$). The structure of a fat is shown in the diagram opposite.

ether (*n*) an organic (p.55) compound with two hydrocarbon (p.172) groups, either alkyl (p.180) or aryl (p.180) joined to one oxygen atom. Ethers are very flammable (p.21) but otherwise are unreactive (p.62).

alkyl halide an organic (p.55) compound formed when one hydrogen atom of an alkane (p.172) is replaced (p.68) by a halogen (p.117) atom. They are named from the halogen and the alkane, e.g. chloroethane, bromopropane, iodoethane. Alkyl chlorides are prepared from alcohols (p.175) by the action of sulphur dichloride oxide (thionyl chloride, $SOCl_2$). They are very reactive, and used in many organic preparations (p.43).

amide (*n*) an organic (p.55) compound containing the functional group $- CONH_2$, e.g. ethanamide CH_3CONH_2. Amides are named from the corresponding carboxylic acid (↑), e.g. propanoic acid forms propanamide. They are formed by dehydrating (p.66) the ammonium salt of the corresponding carboxylic acid.

amine (*n*) an organic (p.55) compound containing the functional group − NH_2. Amines are formed by the reaction between ammonia and an alkyl halide (p.177). The amines can be classified by the number of hydrogen atoms in ammonia that are replaced by alkyl (p.180) or aryl groups, e.g.

Primary amines $R-NH_2$ methylamine CH_3NH_2
Secondary amines R_2-NH dimethylamine $(CH_3)_2NH$
Tertiary amines R_3-N trimethylamine $(CH_3)_3N$

The amines are colourless (p.15) gases or liquids with a strong fishy odour; they are soluble in water forming weak bases, which form salts with inorganic (p.55) acids, e.g. methylamine reacts with hydrochloric acid to form methylammonium chloride $(CH_3NH_3)^+ Cl^-$.

nitrile (*n*) an organic (p.55) compound containing the functional group − CN. The nitriles are named after the corresponding (p.233) alkane with the same number of carbon atoms, e.g. CH_3CH_2CN is propanonitrile. Nitriles are prepared by the action of sodium cyanide (NaCN) on an alkyl halide (p.177). They undergo (p.213) hydrolysis forming first an amide (p.177), then a carboxylic acid (p.176); they undergo reduction (p.193) to an amine (↑).

isocyanide (*n*) an organic (p.55) compound containing the group − NC; they are given names such as isocyanoethane (C_2H_5NC). Isocyanides are prepared by the action of silver cyanide (AgCN) on the corresponding (p.233) alkyl halide (p.177). They are colourless (p.15) liquids with a bad odour and are highly poisonous. Isocyanides are hydrolyzed (p.190) to primary amines and reduced (p.193) to secondary amines.

amino acid a carboxylic acid (p.176) with one hydrogen atom replaced by an amino group (− NH_2), e.g. NH_2CH_2COOH, aminoethanoic acid, with a trivial name (p.44) of glycine. Amino acids in solution form both cations (p.125) and anions (p.125) and thus are amphoteric (p.46) in their reactions; they behave as carboxylic acids and as amines. Amino acids are the constituents (p.54) of proteins (p.209); they are capable of forming long chains by condensation (p.191).

amine

methylamine
primary amine

formula CH_3NH_2

nitrile

ethanenitrile

formula CH_3CN

● carbon atom
● hydrogen atom
◐ oxygen atom
◓ nitrogen atom

isocyanide

methyl isocyanide

CH_3NC formula

amino acid
aminoethanoic acid
(glycine)

NH_2CH_2COOH formula

benzene

formula for benzene or

ethyl benzene
a homologue with
a straight chain

CH₂ CH₃

straight chain

aromatic compound

naphthalene

aromatic compound

alicyclic compound
(also aliphatic)

cyclobutane

pyridine

heterocyclic compound

benzene (*n*) a hydrocarbon (p.172) of formula C_6H_6. The six carbon atoms are combined (p.64) in a ring structure, with all six bonds of equal length and equal activity. The bonds are neither single bonds (p.181) nor double bonds, but have a chemical character peculiar to themselves; this property is called **aromaticity**. Benzene is the first member of two homologous series (p.172); in one, chains (p.182) of carbon atoms are joined to the benzene ring; in the other, two or more benzene rings are combined. The formula of benzene is shown as a hexagon with a circle inside it; the circle represents 6 electrons which do not take part in any one bond, and are known as **delocalized electrons**. The delocalized electrons give benzene its aromatic character. Benzene is a colourless (p.15) liquid with a pleasant (aromatic) smell, is flammable (p.21), a good solvent for organic compounds and is chemically unreactive.

naphthalene (*n*) a hydrocarbon (p.172) of formula $C_{10}H_8$. The ten carbon atoms are combined (p.64) in two benzene (↑) ring structures. Naphthalene is a white solid with a strong odour; it is a major (p.226) constituent (p.54) of coal tar and is chemically reactive (p.62). It is a homologue (p.172) of benzene.

aromatic (*adj*) describes a compound containing (p.55) a benzene (↑) ring. The compound has chemical properties similar to those of benzene.

aliphatic (*adj*) describes a compound consisting of straight chains (p.182) and branched chains, e.g. as in the alkanes, alkenes, and alkynes. Aliphatic compounds include alicyclic (↓) compounds.

alicyclic (*adj*) describes a compound consisting of single bonds, or double bonds, with the carbon atoms combined in a ring. Such compounds are aliphatic (↑) and not aromatic (↑), e.g. cyclobutane is an alicyclic compound.

heterocyclic (*adj*) describes a compound with an aromatic (↑) or aliphatic (↑) ring structure which includes at least one atom that is not carbon, e.g. pyridine C_5H_5N.

alkyl (*adj*) describes an organic (p.55) group formed when one atom of hydrogen is taken from an alkane (p.172); examples of alkyl groups are: methyl (CH_3); ethyl (C_2H_5); pentyl (C_5H_{11}). The general formula of an alkyl group is C_nH_{2n+1}. **alkylation** (*n*).

acyl (*adj*) describes an organic (p.55) group formed when a hydroxyl group ($-OH$) is taken from a carboxylic acid (p.176). *See acyl chloride (p.176).*

aryl (*adj*) an aromatic (p.179) group formed when one atom of hydrogen is taken from an aromatic hydrocarbon (p.172). **arylation** (*n*).

phenol (*n*) an aromatic (p.179) compound of formula C_6H_5OH. It is a white crystalline (p.15) solid with a characteristic odour (p.15); it reacts (p.62) as a weak acid (p.45), forming **phenates**. Phenol is obtained from coal tar; it is used as a common antiseptic and is also important in the plastics (p.210) industry. Homologues (p.172) of phenol are called **phenols**.

phenate
OH + NaOH → ONa + H_2O
sodium phenate

nitro compound a combination of an aryl (↑) group and a nitro group ($-NO_2$) (p.186). Nitro compounds are important because (a) aromatic (p.179) hydrocarbons are easily nitrated by concentrated nitric acid, and (b) they are chemically reactive (p.62). Both nitro and dinitro compounds can be prepared. Alkyl nitro compounds also exist (p.213) but are less important.

diazonium salt a combination of an aryl (↑) group with the azo group $- N^+ \equiv N$ which forms a salt with an inorganic anion (p.125), e.g. $(C_6H_5 - N^+ \equiv N)\ Cl^-$ which is benzenediazonium chloride. Diazonium salts are prepared from primary aromatic (p.179) amines by the action of nitrous acid. The salts are unstable unless kept at a temperature below 0°C. They are important because many dyes (p.162) are manufactured from diazonium salts.

OH

phenol

formula

nitrobenzene

NO_2

formula

nitro compound

● carbon atom
● hydrogen atom
○ oxygen atom
● nitrogen atom
○ chlorine atom

diazonium salt

nitro compound

formula

$N^+ \equiv NCl^-$

benzene diazonium chloride

C:H:O = 1:2:1
by mass
empirical formula
CH$_2$O
for ethanoic acid

relative molecular
mass for ethanoic
acid is 60

molecular formula C$_2$H$_4$O$_2$

structural formula
CH$_3$COOH

● carbon atom
● hydrogen atom
○ oxygen atom

ethanoic acid

graphic formula

empirical formula a formula (p.78) which gives the simplest ratio of elements in a compound, e.g. the empirical formula of ethanoic acid is CH$_2$O.

molecular formula a formula (p.78) which shows the number of atoms of each element in a molecule of a compound. This formula is found from the relative molecular mass of a compound. For example, CH$_2$O is the empirical formula (↑) of ethanoic acid and of glucose (p.206). Their respective (p.233) relative molecular masses are 60 and 180. The relative molecular mass of CH$_2$O is 30, hence ethanoic acid is C$_2$H$_4$O$_2$ and glucose is C$_6$H$_{12}$O$_6$. These are molecular formulae.

structural formula a formula (p.78) which shows how the atoms in a molecule are grouped together, e.g. the structural formula of ethanoic acid is CH$_3$COOH. A structural formula of an organic (p.55) compound indicates (p.38) how the compound will react.

graphic formula a formula (p.78) which shows how the atoms in a molecule are oriented (p.93) in space, relative (p.232) to each other.

general formula a formula (p.78) which allows the formula for any member of an homologous series (p.172) to be written, e.g. C$_n$H$_{2n}$ is the general formula for alkenes. Putting n = 2,3,4, etc. gives the formula for members of the series.

single bond a covalent bond (p.136), it is oriented (p.93) in space and allows atoms to turn relative (p.232) to each other.

double bond two covalent bonds (p.136) joining together two atoms; each covalent bond is formed by two atoms sharing two electrons. The double bond between two carbon atoms is important, it prevents the atoms turning about the bond and also can cause geo-metrical isomerism (p.184). A double bond makes an organic (p.55) compound reactive (p.62).

triple bond three covalent bonds (p.136) joining together two atoms; each covalent bond is formed by two atoms sharing two electrons. The triple bond between two carbon atoms is important. It has the same effect as a double bond (↑); but is much more reactive.

a straight carbon chain

straight chain

formula of a straight chain

chain (*n*) a structure in which each carbon atom is joined to the next carbon atom to form a chain, or line, of atoms. A chain may be straight (↓) or branched (↓).

straight chain a chain of carbon atoms in which any one carbon atom is not joined to more than two other carbon atoms, e.g. as in hexane.

branched chain a chain of carbon atoms in which one or more carbon atoms may be joined to 3 or 4 other carbon atoms. *See diagram.*

cyclic chain a chain of carbon atoms forming a circle in which the carbon atoms may be joined by either single or double bonds. The resulting (p.39) compound has an aliphatic (p.179) nature.

ring chain a chain of carbon atoms which can be cyclic (↑) or can be a benzene (p.179) ring with neither single nor double bonds.

isomerism (*n*) the property of having the same molecular formula (p.181), but different structural formulae, e.g. ethanol and methoxymethane (dimethyl ether) have the same molecular formula, but their structural formulae are CH_3CH_2OH and $CH_3O\ CH_3$ respectively (p.233) There are different kinds of isomerism Isomerism is very common in organic (p.55) compounds. **isomer** *(n)*, **isomeric** *(adj)*.

isomer (*n*) one of two, or more, compounds which exhibit (p.221) isomerism (↑), e.g. ethanol and methoxymethane are isomers.

structural isomerism isomerism (↑) exhibited by two, or more, organic (p.55) compounds which have the same molecular formulae, but their different structures give them different physical or chemical properties, e.g. ethanol and methoxymethane are structural isomers; a branched chain and a straight chain alkane (p.172) with the same number of carbon atoms are isomers of each other. *See diagram.*

branched carbon chain

branched chain

formula for a branched carbon chain

structural isomerism
ethanol

CH_3CH_2OH

methoxymethane

$CH_3 - O - CH_3$

two compounds exhibiting structural isomerism

butane

two compounds exhibiting structural isomerism

methyl propane

isomerism

OH
|
HOOC—C—H
|
HOOC—C—H
|
OH

right rotation

OH
|
H—C—COOH
|
H—C—COOH
|
OH

left rotation

2, 3 dihydroxybutanedioic acid

tartaric acid

(two asymmetric
carbon atoms in
the molecule) C

stereoisomerism (*n*) isomerism that results from a different arrangement in space of the atoms in a molecule. The structural formula does not show the different arrangements which are possible, but the graphic formula of such compounds can show the difference. Stereoisomerism results from the tetrahedral (p.83) directions of the four covalent bonds of carbon. There are two kinds of stereoisomerism: optical isomerism (↓) and geometrical isomerism (p.184).

optical isomerism optical isomerism exists when a molecule is not identical (p.233) with its image in a mirror. Two spatial (p.211) arrangements of the atoms are thus possible. The two structures are chemically identical, but one structure rotates (p.218) the plane of polarized light to the left and the other structure rotates it to the right; otherwise their physical properties are the same, e.g. lactic acid exhibits (p.221) optical isomerism.

enantiomorph (*n*) (1) a molecule exhibiting (p.221) optical isomerism (↑) which is one of two or more isomeric (↑) forms. (2) one of two crystalline forms when each crystal is the mirror image of the other. **enantiomorphic** (*adj*).

●C
◉CH₃
●H
■COOH
○OH

bonds
in plane ——
in front - - -
of plane
behind· · · · ·
plane

optical
isomerism

left
rotating

right
rotating

2-hydroxy
propanoic acid

CH₃CH(OH)COOH

optically
active forms
of lactic acid

enantiomorph

CH₃
|
H—C—OH
|
COOH

C asymmetric
carbon atom

lactic acid

asymmetric carbon atom a carbon atom which has four different atoms or groups attached to it. It is the most common cause of optical isomerism (↑), e.g. lactic acid.

racemate (*n*) a physical mixture of two enantiomorphs (↑), one with rotation of the plane of polarized light to the right, and the other to the left. The result is optically inactive. When optically active compounds are synthesized, a racemate is produced. **racemic** (*adj*).

geometrical isomerism geometrical isomerism results from two conditions: (1) two carbon atoms joined by a double bond; (2) each carbon atom must have two different atoms or groups joined to it. A double bond prevents rotation (p.218) of the carbon atoms, so two spatial (p.211) arrangements of the atoms are possible. Geometrical isomers have very different physical properties, e.g. but-2-ene has one isomer with a melting point of −139°C and the other −106°C. Their chemical properties tend to be the same, but certain differences also exist, e.g. the difference between maleic and fumaric acids, which are isomers. **geometrical isomer** (*n*).

geometrical isomers

cis-configuration

trans-configuration

- carbon atom
- hydrogen atom
○ chlorine atom

bonds

in plane ——
in front ----
of plane
behind ·····
plane

CICH = CHCI

1, 2 dichloroethene

cis-configuration (*n*) a geometrical isomer (↑) with two like groups on the same side of the double bond. *See diagram.*

trans-configuration (*n*) a geometrical isomer (↑) with two like groups on opposite sides of the double bond. *See diagram.*

tautomerism (*n*) a state of dynamic equilibrium (p.150) between spontaneously convertible (p.73) isomers. Generally in the conversion of one isomer to another, a hydrogen atom changes its place in the molecular structure, e.g. as in ethyl 3-oxobutanoate (ethyl acetoacetate). The two isomeric forms in tautomerism are the 'keto' and 'enol' forms, shown in the diagram opposite.

tautomer (*n*) an individual isomer of tautomerism (↑); the 'keto' or 'enol' form of the tautomeric compound. **tautomeric** (*adj*).

methyl (*adj*) the alkyl (p.180) group $-CH_3$.
ethyl (*adj*) the alkyl (p.180) group $-C_2H_5$.
propyl (*adj*) the alkyl (p.180) group $-C_3H_7$.
butyl (*adj*) the alkyl (p.180) group $-C_4H_9$.
pentyl (*adj*) the alkyl (p.180) group $-C_5H_{11}$.

tautomerism

ethyl 3 oxobutanate
(ethyl acetoacetate)

addition reaction

hydroxyl group

dehydration

an aliphatic reaction

an aromatic reaction

$$ArOH = ArO^- + H^+$$

carbonyl group

$$>C = O$$

saturated[2] (*adj*) describes a carbon chain in which there are only single bonds connecting individual carbon atoms. A saturated chain is unreactive, as alkanes (p.172) are. **saturation** (*n*).

unsaturated[2] (*adj*) describes a carbon chain in which there is at least one double or triple bond. An unsaturated chain is reactive, particularly for addition reactions (p.188).

polyunsaturated (*adj*) describes a carboxylic acid containing more than two double bonds.

functional group an atom, or group of atoms, which gives an organic (p.55) compound its reactivity. A functional group is joined to an alkyl (p.180) or aryl group, which forms the homologous series (p.172) and the functional group gives the chemical characteristics. A long carbon chain with many carbon atoms reduces (p.219) the chemical activity of the functional group. Examples of functional groups are: hydroxyl, carboxyl, amino, chloro. In general formulae an alkyl group is represented by R and an aryl group by Ar or Ph, e.g. RCOOH is a general formula for a carboxylic acid; $ArNO_2$ is an aromatic (p.179) nitro compound.

hydroxyl group the functional group ($\uparrow$) $-OH$; it occurs (p.154) in alcohols (p.175) and phenols (p.180). Combined with alkyl groups, the hydroxyl group reacts with the hydrogen of acids to form water. Combined with aryl groups, the hydrogen of the hydroxyl group forms hydrogen ions, making the phenols weakly acidic.

monohydric (*adj*) describes an alcohol with one hydroxyl group. *See alcohol (p.175).*

dihydric (*adj*) describes an alcohol with two hydroxyl groups. *See alcohol (p.175).*

trihydric (*adj*) describes an alcohol with three hydroxyl groups. *See alcohol (p.175).*

carbonyl group the functional group ($\uparrow$) $= CO$; it occurs in aldehydes (p.175) and ketones (p.176). The carbonyl group has a double bond (p.181) between the carbon and oxygen. This double bond undergoes (p.213) addition reactions (p.188). A hydrogen atom joined to the carbon atom of the carbonyl group makes it more reactive, as in aldehydes.

carboxyl group the functional group (p.185) −COOH. It forms weak acids and by neutralization it forms salts, e.g. sodium ethanoate. With alcohols (p.175) it forms esters (p.177).

carboxyl group   ionization

sulphonate group the functional group (p.185) −SO$_2$.OH. It occurs almost entirely combined with aryl groups (p.180). Aromatic (p.179) sulphonic acids are prepared by the action of concentrated sulphuric acid on benzene (p.179) and its homologues; they form salts with alkalis, and the salts are used as detergents (p.171).

nitro group the functional group (p.185) −NO$_2$. Aliphatic (p.179) nitro compounds exhibit (p.221) tautomerism (p.184). They are of less importance than the aromatic (p.179) nitro compounds. Benzene and its homologues are converted to nitro compounds by nitration (p.193). Aromatic nitro compounds are reduced (p.70) to amines in several stages; some intermediate compounds are useful industrially. *See equations opposite.*

tautomers of aliphatic nitro compounds

nitro compound

reduction of aromatic nitro compound

phenyl hydroxylamine

amino group the functional group (p.185) −NH$_2$. It occurs in primary amines, both aliphatic (p.179) and aromatic; in the amides as part of the amido (↓) group; and in the amino acids. The amino group causes amines and amino acids to have a basic (p.46) reaction.

amido group the functional group (p.185) −CONH$_2$. It occurs in amides and causes them to be amphoteric (p.46) in nature. The group is dehydrated to form a nitrile (p.178).

azo group the functional group (p.185) −N=N−. It occurs in diazo compounds (p.180) and is a highly reactive group.

cyano group the functional group (p.185) −C≡N. It occurs in the nitriles (p.178). The cyano group is hydrolyzed (p.66) to the carboxyl group and reduced (p.70) to a primary amine

phenyl azoxy benzene

phenyl azobenzene

diphenyl-hydrazine

phenylamine (aniline)

chloro group the functional atom (p.185) —Cl. It occurs in the alkyl chlorides (p.177) and the aromatic chloro compounds. The alkyl chlorides are reactive, taking part in substitution (p.188) and elimination (p.189) reactions. The aromatic chloro compounds are relatively (p.232) inert (p.19). The presence of a nitro group (p.186) in the aromatic hydrocarbon increases the reactivity of the chloro group.

bromo group the functional atom (p.185) —Br. It occurs in the alkyl bromides, e.g. bromoethane CH_3CH_2Br, and in the aromatic (p.179) bromo compounds. The alkyl and aromatic bromo compounds are more reactive than the corresponding (p.233) chloro compounds, e.g. bromobenzene is more reactive than chlorobenzene.

bromoethane
CH_3CH_2

iodo group the functional atom (p.185) —I. It occurs in the alkyl (p.180) iodides, e.g. iodoethane CH_3CH_2I, and in the aromatic (p.179) iodo compounds, e.g. iodobenzene C_6H_5I. The alkyl and aromatic iodo compounds are more reactive than the corresponding bromo compounds (↑).

iodobenzene

chromophore (n) any chemical group which causes a compound to have a distinctive (p.224) colour. In synthetic (p.200) organic dyes (p.162), such groups as the azo group cause the colour of a compound. **chromophoric** (adj).

auxochrome (n) a chemical group which deepens the colour of a compound, where red is taken as the deepest colour of the spectrum and violet as the least deep, e.g. the addition of an auxochrome can change a blue substance to a green substance. In addition, an auxochrome intensifies (p.230) the colour. The amino group (↑) is an example of an auxochrome. **auxochromic** (adj).

auxochrome

indigo

=O chromophores
>NH auxochromes

leuco base a compound formed from a dye (p.162) by reduction (p.70); it is colourless. An insoluble dye is converted to a leuco base which is soluble in alkalis; the alkaline solution is used for dyeing and the dye colour returned by oxidation, e.g. indigo is converted to a leuco base and the colour returned by subsequent oxidation.

leuco compound a name for leuco base (↑).

a $CH_3CH_2Br + NaOH \rightarrow CH_3CH_2OH + NaBr$

b $+ HO\ NO_2 \rightarrow$ $+ HOH$

c $CH_3COOH \rightarrow CH_3COOAg$

$CH_3COOAg + Br_2 \rightarrow CH_3Br + CO_2 + AgBr$

substitution (n) a process in which an atom or a functional group in an organic (p.55) compound is replaced, directly or indirectly, by another atom or functional group. Substitution is one of the most important organic reactions. Examples of substitution are: (a) substituting a hydroxyl group for a halogen in an aliphatic (p.179) compound; (b) substituting a nitro group (p.186) for a hydrogen atom in benzene (p.179); (c) substituting a bromine atom for a carboxyl group in an aliphatic compound. **substitute** (v), **substitution** (adj).

addition (n) a process in which two substances react to form only one substance. In organic reactions, addition usually takes place across a double bond (p.181), e.g. hydrogen chloride adds across the double bond of ethene. **addition** (adj).

addition   hydrogenation

$$CH_2 \quad H \quad CH_3 \qquad CH_2 \qquad CH_3$$
$$\| \quad + \quad | \rightarrow | \qquad \| \quad + H_2 \rightarrow |$$
$$CH_2 \quad Cl \quad CH_2Cl \qquad CH_2 \qquad CH_3$$

hydrogenation (n) an addition process (↑) in which hydrogen is added to a molecule in the presence of a suitable catalyst (often finely divided nickel) and at a suitable temperature and pressure. Hydrogenation converts unsaturated (p.185) compounds into saturated compounds. **hydrogenate** (v).

alkylation

propene

2, 4 dimethylpentane

2-methylpropane

alkylation (*n*) an addition process (↑) in which an alkane (p.172) undergoes (p.213) addition across the double bond of an alkene (p.173) to form a branched chain (p.182) alkane, e.g. propene and 2-methylpropane form 2,4,dimethylpentane. Alkylation is also the substitution (↑) of an alkyl (p.180) group in place of a hydrogen atom. *See Friedel-Crafts reaction (p.199).*

elimination (*n*) the removal (p.215) of atoms from a molecule; this usually results in the formation of a double bond, e.g. the removal of the elements of hydrogen and a halide (p.50) from an alkyl halide (p.177) to form an alkene. This is done by the action of a hot, concentrated solution of potassium hydroxide in ethanol (called alcoholic potash), e.g. bromoethane is converted to ethene. Other types of elimination include dehydrogenation (p.190) and dehydration (p.190).

elimination

$$CH_3 \xrightarrow[\text{potash}]{\text{alcoholic}} CH_2 + (HBr)$$
$$CH_2Br \qquad\qquad CH_2$$

dehydrogenation (*n*) the removal (p.215) of combined hydrogen from an organic (p.55) compound; the process uses a suitable catalyst (e.g. aluminium and chromium oxides) at a high temperature (e.g. 600°C). For example, butane is dehydrogenated to butene. This is an elimination (p.189) process. **dehydrogenate** (*v*).

dehydration (*n*) the removal (p.215) of the elements of water from an organic (p.55) compound. For example, (a) the action of excess concentrated sulphuric acid on ethanol produces ethoxyethane (diethyl ether); (b) the action of phosphorus pentoxide on ethanamide produces ethanenitrile. This is an elimination (p.189) process. **dehydrate** (*v*).

hydrolysis[2] (*n*) the reverse process of esterification (↓) and of condensation (↓), and also the substitution of a hydroxyl group into a compound. Hydrolysis of esters is catalyzed by hydrogen ions in acids or by hydroxyl ions in alkalis. Condensation products can be hydrolyzed by boiling with dilute acids. Water alone hydrolyzes alkyl halides (p.177). **hydrolyze** (*v*).

condensation[3] (*n*) a reaction in which two
↳organic (p.55) compounds combine to form
one compound with the elimination of water,
hydrogen chloride or ammonia. The most
common compound eliminated is water.
Examples of condensation are (a) aldehydes
and hydroxylamine condensing to aldoximes,
e.g. ethanal oxime $CH_3CH=NOH$; (b)
polymerization (p.207) with condensation, as in
the urea-methanal condensation
polymerization.

urea

methanal

polymer
unit

$$O=C + O=C \rightarrow C=O$$

condensation

esterification (*n*) the process of converting an
acid into an ester (p.177). The reaction is
reversible, so the water formed in the reaction
is removed by concentrated sulphuric acid to
move the equilibrium mixture (p.150) in the
direction of the ester. For example, the
esterification of ethanoic acid with methanol:
$CH_3COOH + H\ CH_2OH \rightleftharpoons CH_3COOCH_3 + H_2O$
This is a form of condensation (↑).

$$CH_3COOCH_3 + NaOH \rightarrow CH_3OH + CH_3COONa$$

methyl ethanoate → methanol + sodium ethanoate

saponification

saponification (*n*) hydrolysis (p.190) of an ester using boiling aqueous sodium or potassium hydroxide. The alkali combines with and removes (p.215) the acid from the products, and shifts the equilibrium mixture (p.150) towards the hydrolyzed product. **saponify** (*v*).

halogenation (*n*) the process of substituting (p.188) a halogen (p.117) atom for a hydrogen atom in a molecule of an organic (p.55) compound, e.g. the halogenation of phenol (p.180).

chlorination (*n*) halogenation (↑) when chlorine is the halogen. Chlorinating agents for aliphatic (p.179) compounds include phosphorus penta-chloride, phosphorus trichloride and sulphur dichloride oxide (thionyl chloride, $SOCl_2$). For aromatic (p.179) compounds, a 'halogen car-rier', such as aluminium chloride or powdered iron, is used and chlorine is passed through at room temperature, causing substitution in the benzene ring (p.179). **chlorinate** (*v*).

$$CH_3CH_2COOH + Cl_2 \rightarrow CH_3CHClCOOH + HCl$$

propanoic acid   2-chloro propanoic acid

chlorination

bromination

$$CH_3 \, CO \, CH_3 + Br_2 \rightarrow CH_3COCH_2Br + HBr$$

propanone   bromopropanone

bromination (*n*) halogenation (↑) when bromine is the halogen. For aliphatic (p.179) compounds, red phosphorus and bromine are used as brominating agents. For aromatic (p.179) compounds, a 'halogen carrier' is used as for chlorination (↑). **brominate** (*v*).

iodination (*n*) halogenation (↑) when iodine is the halogen. For aliphatic (p.179) compounds, red phosphorus and iodine are used as iodinating agents. For aromatic (p.179) compounds, iodine in the presence of mercury (II) oxide is used.

$$2\,\text{C}_6\text{H}_6 + \text{I}_2 \rightarrow 2\,\text{C}_6\text{H}_5\text{I}$$

iodination

reduction[2] (*n*) the reduction of organic (p.55) compounds is either hydrogenation (p.188) or the removal of oxygen, e.g. the reduction of ethanoic acid to ethanol or to ethane.

reduction

$$\text{CH}_3\text{COOH} \xrightarrow{\text{LiAlH}_4} \text{CH}_3\text{CH}_2\text{OH}$$

$$\text{CH}_3\text{COOH} \xrightarrow{\text{HI}} \text{CH}_3\,\text{CH}_3$$

reduction (hydrogenation)

$$\begin{array}{c} \text{CH}_2 \\ \| \\ \text{CH}_2 \end{array} + \text{H}_2 \rightarrow \begin{array}{c} \text{CH}_3 \\ | \\ \text{CH}_3 \end{array}$$

$$\text{C}_6\text{H}_6 + \text{HO NO}_2$$

↓

$$\text{C}_6\text{H}_5\text{NO}_2 + \text{H}_2\text{O}$$

nitration

nitration (*n*) the substitution (p.188) of a nitro group (p.186) in an organic (p.55) compound, using concentrated nitric acid. With aliphatic (p.179) compounds, a high temperature is needed. With aromatic (p.179) compounds, the reaction takes place at much lower temperatures; concentrated sulphuric acid is used to quicken the reaction.

sulphonation (*n*) the substitution (p.188) of a sulphonate group (p.186) in place of a hydrogen atom in an organic (p.55) compound. Aromatic (p.179) compounds undergo (p.213) sulphonation readily, but aliphatic (p.179) hydrocarbons are extremely difficult to sulphonate. **sulphonate** (*v*).

sulphonation

benzene

↓ oleum

SO₂OH

benzenesulphonic acid

$$C_2H_5COOH + CH_2N_2 \rightarrow C_2H_5COOCH_3 + N_2$$

diazomethane   **methylation**

methylation (*n*) a process in which a methyl group (p.184) is substituted for a hydrogen atom in an organic compound. Methylating agents include dimethyl sulphate, diazomethane and iodomethane. **methylate** (*v*).

cracking (*n*) an industrial (p.157) process in which hydrocarbons from petroleum fractions (p.160) are strongly heated under pressure; large molecules are broken up into smaller molecules, e.g. the higher fractions (p.202) are converted to petrol and kerosene. **crack** (*v*).

fermentation (*n*) the decomposition of carbohydrates (p.205) caused by enzymatic (p.72) action; the enzymes are produced by yeasts or by bacteria. **ferment** (*v*).

ortho-para-directing (*adj*) a functional group (p.185) in a benzene ring directs (p.233) the position of further substitution (p.188) in the ring. Some functional groups direct towards the *ortho* and *para* positions, and a second substitution will take place in these positions. An *ortho-para*-directing group activates (p.21) the benzene ring, so the second substitution is faster. The following groups are *ortho-para*-directing: alkyl, amino, hydroxyl, halogeno, methoxy.

meta-directing (*adj*) a functional group in a benzene ring directs (p.233) the position of further substitution in the ring. See *ortho-para* directing (↑). Some groups direct towards the *meta* position, these reduce (p.219) the activity of the ring so that the second substitution is slower and more difficult. The following groups are *meta*-directing: nitro, sulphonate, carboxyl, aldehyde, cyano.

ortho-para-directing

$$1 \quad CH_3\ CH = CH_2 \rightarrow CH_3\ CH^{\oplus} - CH_2^{\ominus}$$

the mechanism of a reaction

product
2-chloropropane

$$2 \quad \begin{array}{c} CH_3 \\ | \\ HC^{\oplus} \\ | \\ H - C \\ | \\ H^{\ominus} \end{array} + \begin{array}{c} Cl^{\ominus} \\ H^{\oplus} \end{array} \rightarrow \begin{array}{c} CH_3 \\ | \\ HC - Cl \\ | \\ HC - H \\ | \\ H \end{array}$$

mechanism

mechanism (*n*) the mechanism of a reaction gives an explanation of the way in which a reaction takes place. Reactivity occurs (p.63) because points of electron excess (p.230) or electron deficiency (p.232) appear in an organic (p.55) molecule and chemical attack takes place at these points. The reaction mechanism of the addition of hydrogen chloride to propene, shown above, explains how the reaction takes place and why the particular product is obtained.

ozonolysis (*n*) the process of passing ozonized oxygen ($O_2 + O_3$) through a solution of an alkene containing zinc dust and ethanoic acid. An ozonide, *see diagram*, is first formed by ozone adding across the double bond. This decomposes to form two compounds with a carbonyl group (p.185); these compounds may be aldehydes or ketones. Ozonolysis is used to determine (p.222) the position of a double bond in an unsaturated (p.185) molecule, by identifying (p.225) the products.

$$CH_3\ CH = CHC_2H_5 + O_3 \rightarrow CH_3\ CH - CH\ C_2H_5$$

ozonide

$$R\ CH = C \begin{array}{c} R^1 \\ R^2 \end{array}$$

ozonide $\longrightarrow$ $$R\ C = O \ + \ R^1 - C = O$$
$$\begin{array}{cc} | & | \\ H & R^2 \end{array}$$

ozonolysis   aldehyde   ketone

Schiff's reagent red fuchsine dye decolorized (p.73) by sulphur dioxide. When an aldehyde is added to Schiff's reagent, the red, or pink, colour returns. Ketones have no effect on the reagent, so the two kinds of compound can be distinguished (p.224).

Fehling's test a solution containing copper (II) sulphate, sodium potassium tartrate and sodium hydroxide is added to a solution to test for a sugar (p.205). If a red precipitate appears when the solution is boiled, a reducing sugar (p.206) is present (p.217). A red precipitate also indicates that an aldehyde (p.175) is present.

Benedict's test a solution containing copper (II) sulphate, sodium citrate and sodium carbonate is added to a solution to test for a sugar (p.205). If a red precipitate appears when the solution is boiled, a reducing sugar (p.206) is present (p.217). A red precipitate also indicates that an aldehyde (p.175) is present.

biuret test sodium hydroxide solution is added to a substance, followed by one or two drops of 1% copper (II) sulphate solution. A violet colour shows that protein (p.209) is present.

iodine test iodine solution is added to a substance. A deep blue colour indicates that starch (p.207) is present.

Lassaigne test a test to identify (p.225) the elements in an organic compound. The compound is fused with sodium metal in a test-tube. The tube is then broken and an aqueous solution of its contents formed. Carbon and nitrogen form sodium cyanide, tested for by iron (II) sulphate. Sulphur forms sodium sulphide, tested for as hydrogen sulphide. A halogen forms a sodium halide, tested for by silver nitrate. Thus C, N, S, Cl, Br, I are tested for.

$CH_3 \cdot CH_2 \cdot CH_2 \cdot Br$

$+$ $+ \, 2Na \longrightarrow CH_3CH_2CH_2CH_2CH_2CH_3 \; + \; 2NaBr$

$CH_3 \cdot CH_2 \cdot CH_2 \cdot Br$ hexane

bromopropane   **Wurtz reaction**

Fittig reaction

Wurtz reaction the production of an alkane (p.172) from an alkyl halide (p.177) by the action of metallic sodium, using an inert (p.19) solvent, e.g.

$$2C_2H_5Br + 2Na \rightarrow C_2H_5-C_2H_5 + 2NaBr$$

The general reaction is:

$$2RX + 2Na \rightarrow R\text{-}R + 2NaX.$$

Fittig reaction the production of a benzene homologue (p.172) from an aromatic (p.179) halide by the action of metallic sodium in ether.

Wurtz-Fittig reaction the production of an alkyl benzene from an aliphatic halide and an aromatic halide by the action of metallic sodium in ether. By-products are also formed.

Wurtz-Fittig reaction

Ullmann reaction

Ullmann reaction the production of higher aromatic (p.179) homologues from a bromo or iodo compound by the action of copper powder. *See equation below.* The general reaction is $2ArX \rightarrow Ar-Ar$ where $\bar{A}r$ is an aryl (p.180) group and X is iodine or bromine.

Williamson's synthesis a process for preparing simple or mixed ethers (p.177) from an alkyl halide (p.177) and the sodium derivative (p.200) of an alcohol, e.g.
$$C_2H_5I + CH_3ONa \rightarrow C_2H_5 - O - CH_3 + NaI$$
Methoxyethane ($CH_3OC_2H_5$) is a mixed ether.

Grignard reagent a reagent which is prepared by the action of metallic magnesium on an alkyl halide in ether. An alkyl magnesium halide is formed, and this is the reagent, e.g.
$$CH_3CH_2Br + Mg \rightarrow CH_3CH_2MgBr$$
A Grignard reagent undergoes (p.213) many different reactions and the different reagents that can be prepared are used in the synthesis of many organic compounds. The symbol R is used in a Grignard reagent to represent any alkyl group (p.180) and the symbol X to represent a halogen (p.117). Grignard reagents can also be prepared using aryl (p.180) halides, e.g. $C_6H_5Br + Mg \rightarrow C_6H_5MgBr$
Only bromo and iodo compounds can be used for aryl halides.

Sandmeyer reaction the use of copper (I) chloride in concentrated hydrochloric acid to convert an aqueous diazonium (p.180) chloride to a chloro-compound. A similar reaction takes place with copper (I) bromide and a diazonium bromide. A copper salt is unnecessary for the corresponding iodo compound.

Gattermann reaction[1] the use of copper powder to convert a diazonium salt to a halogeno compound, e.g. $C_6H_5 - N^+ \equiv N\ Cl^- \rightarrow C_6H_5Cl$
The yield (p.159) from this reaction is not as good as for the Sandmeyer reaction (↑).

Sandmeyer reaction

Friedel-Crafts reaction

Friedel-Crafts reaction a process for the alkylation (p.189) of benzene rings (p.179). Alkyl halides (p.177) react with aromatic hydrocarbons in the presence of anhydrous aluminium chloride.

Reimer-Tiemann reaction a reaction between phenols (p.180), aqueous sodium hydroxide and trichloromethane (chloroform), which puts an aldehyde group (p.175) on the benzene ring. The reaction does not produce a high yield (p.159) but is the easiest method of putting an aldehyde group on a benzene ring The substitution generally takes place in the *ortho*-position.

Reimer-Tiemann reaction

Gattermann reaction[2] the combination of phenols with hydrogen chloride or hydrogen cyanide to give a product which is hydrolyzed by water to an aldehyde.

Gattermann reaction

Cannizzaro reaction aldehydes (p.175) with no hydrogen atom on the second carbon atom in the aliphatic group, when treated (p.38) with cold concentrated aqueous sodium hydroxide, are both oxidized and reduced (p.70) at the same time; e.g. with methanal the reaction is:

$$2HCHO + NaOH \rightarrow HCH_2OH + HCOONa$$

The reaction also takes place with aromatic (p.179) aldehydes.

Cannizzaro reaction

$$2H_3C-\underset{\underset{CH_3}{|}}{\overset{\overset{CH_3}{|}}{C}}-CHO + NaOH \longrightarrow$$

$$H_3C-\underset{\underset{CH_3}{|}}{\overset{\overset{CH_3}{|}}{C}}-CH_2OH + H_3C-\underset{\underset{CH_3}{|}}{\overset{\overset{CH_3}{|}}{C}}-COONa$$

aldol addition a reaction in which aldehydes and ketones form addition compounds across the double bond (p.181) of the carbonyl group (p.185). The original reaction was:

$$CH_3CHO + CH_3CHO \rightleftharpoons$$
$$CH_3.CH(OH)CH_2CHO \text{ (aldol)}$$

The reacting molecules need not be identical. Equilibrium is largely to the left of the reaction, i.e. the addition product is formed only in small quantities. A trace (p.20) of potassium hydroxide catalyzes (p.72) the reaction.

aldol reaction

Kolbe electrolytic reaction a process for preparing alkanes (p.172) by the electrolysis (p.122) of the sodium salts of carboxylic acids. The carboxylic group must be at the end of a carbon chain. A cold concentrated aqueous solution of a salt is electrolyzed using platinum electrodes.

$$CH_3COONa \rightarrow CH_3{-}CH_3 + CO_2 + NaOH$$

Only alkanes with an even number of carbon atoms are prepared by this reaction.

derivative (*n*) a compound made by substitution of one or more atoms, or groups of atoms, in an original substance. Because the compound is a derivative of an original substance, it has the same kind of structure. For example, taking propane as an original substance, then its derivatives include propanol, propanal, propanoic acid and chloropropane. Derivatives are used to determine structure or composition. If the original structure is known, then the structure of the derivative is known. Ozonolysis (p.195) is frequently used to prepare derivatives. The derivatives are identified and from that the original structure of an alkene, including the position of the double bond (p.181), is found. **derive** (*v*).

derivatives

synthesis (*n*) (1) making a compound by chemical means from its elements. (2) making a compound by a series of chemical processes. e.g the synthesis of ascorbic acid (vitamin C). A simple synthesis is shown in the equations opposite. **synthetic** (*adj*), **synthesize** (*v*).

synthetic (*adj*) describes any compound made by synthesis (↑) from simpler compounds. A synthetic substance is one that generally replaces a naturally occurring (p.154) substance.

reforming (*n*) a process in which straight chain alkanes (p.172) or alicyclics (p.179) are dehydrogenated (p.190) and form aromatic or branched chain hydrocarbons.

reflux (*adj*) of a liquid or gas, the flowing back in an opposite direction from its original (p.220) direction. **reflux** (*n*), **reflux** (*v*).

reflux condenser a condenser (p.28) fitted above a flask so that the vapour, formed by heating the flask, is condensed and flows back into the flask. This action prevents (p.216) the flask boiling dry and allows organic reactants sufficient (p.231) time to react while being heated, as the reactants do not escape from the flask.

still (*n*) an apparatus (p.23) used in industry for distillation (p.33). It is generally made of metal and produces (p.62) large quantities of distillate (↓).

distillate (*n*) the liquid obtained as the result of condensation from a process of distillation (p.33).

fractional distillation a process of distillation (p.33) in which the neck of a distillation flask (p.28) is extended (p.213) by a fractionating column (↓). Vapour from the flask condenses and falls back into the flask. Only vapour reaching the top of the column passes to a condenser (p.28) and is collected as a distillate (↑). This process is used to separate liquids which have boiling points close together. A column can have exits at different heights, allowing vapours which condense (p.11) at different temperatures to be separated and collected. This type of fractional distillation is used in industry (p.157), especially in the distillation of petroleum (p.160).

fractionating column an extension (p.213) of the neck of a distillation flask with an exit (p.215) at the top of the column, to pass the vapour to a condenser (p.28). An industrial fractionating column has several exits to take away vapour condensing to a liquid at a particular temperature. *See fractional distillation* (↑).

fraction (*n*) the distillate (p.201) collected at a particular temperature from a fractionating column (p.201), e.g. in the distillation of petroleum, four main fractions are obtained. These fractions can be fractionally distilled again to obtain a better separation (p.34) of the different liquids.

liquid extraction

1   2   3

original solution

solvent

substances dissolved in solvent

gas out

temperature controlled in column

fractions

vapour in

residue of pitch

fractionating column for petroleum

industrial fractional distillation

liquid extraction the removal (p.215) of a substance, dissolved in one solvent, by using a second solvent. The second solvent is added to the original (p.220) solution in a separating funnel (p.27). The two solvents must be immiscible (p.18). The solute (p.86) dissolves in both solvents and the two liquids are separated (p.34) by a separating funnel; e.g. ethanol is extracted from ethoxyethane (diethyl ether) by adding a concentrated solution of sodium chloride, and running off the aqueous layer in which the ethanol is dissolved.

steam distillation

to condenser and distillation

steam

heat

steam distillation steam is blown through the heated mixture of the products of an organic (p.55) reaction. If a liquid product of the reaction is immiscible (p.18) with water, then the vapour of the product together with steam passes into a condenser and the distillate (p.201) contains water and the condensed liquid, e.g. phenylamine (aniline) can be steam distilled. The advantage of steam distillation is that the product is obtained at a temperature below its boiling point.

vacuum distillation distillation is carried out (p.157) at a low pressure. This method (p.221) of distillation is used with substances that decompose (p.65) at temperatures below their boiling points. At a lower pressure, a substance boils at a temperature below its boiling point.

dry distillation

dry distillation the heating of a solid to make it give off (p.41) a vapour (p.11); the vapour is condensed to a liquid. For example, the dry distillation of solid calcium ethanoate (acetate) produces propanone (acetone), which is a volatile (p.18) liquid.

destructive distillation the heating of solid or liquid organic (p.55) materials to a temperature which is high enough to decompose (p.65) the material so that a residue and a distillate (p.201) are produced; a gas may be produced as well. Coal undergoes (p.213) destructive distillation. The residue is coke (p.156); the distillate consists of tar (p.162); and coal gas is also produced. Wood can also be destructively distilled.

destructive distillation of coal

spirit (*n*) (1) a liquid consisting mainly of ethanol (ethyl alcohol) obtained by the distillation (p.33) of fermented (p.194) fruit, potatoes, or cereals. (2) any volatile (p.18) liquid obtained by distillation from natural products such as wood and petroleum; used for combustion (p.58), especially in internal combustion engines. (3) any solvent for fat, gums, paints.

methylated spirit a mixture of ethanol (ethyl alcohol) and methanol (methyl alcohol), usually with added colouring material to show the mixture is poisonous. Methylated spirit is used as a fuel (p.160) and as a solvent (p.86).

absolute alcohol ethanol (ethyl alcohol) from which all traces (p.20) of water have been removed. This is done by adding small quantities of calcium oxide (quicklime) which removes the water from 96% ethanol obtained by distillation.

oils
neutral, viscous, ether soluble immiscible with water

mineral oils
volatile hydrocarbons (alkanes, cycloalkanes)

fixed oils
non-volatile esters of glycerol

essential oils
volatile esters

oil (*n*) (1) any substance, obtained from animals, plants or minerals, which is neutral (p.45), viscous (p.18), combustible (p.58) and soluble in ethanol or ether (ethoxyethane) but insoluble in water. The three main kinds of oil are fixed oils (↓), mineral oils and essential oils (↓). Mineral oils are petroleum (p.160). (2) a neutral fat (p.177) which is liquid below 20°C, is called an oil; such substances are fixed oils (↓). **oily** (*adj*).

essential oil a volatile (p.18) oil produced from a plant and occurring (p.154) especially in the flowers of a plant. Essential oils give a plant its characteristic odour (p.15).

fixed oil a non-volatile oil occurring in plants. Fixed oils are generally edible and are used in cooking, e.g. coconut oil, peanut oil. The fixed oils are esters (p.177) of glycerol (a trihydric alcohol (p.175)) and unsaturated (p.185) or polyunsaturated (p.185) carboxylic acids, although they may contain a small proportion (p.76) of saturated (p.185) carboxylic acids.

hydrocarbon oil any oil obtained industrially (p.157) from petroleum (p.160).

flash point the lowest temperature at which a volatile (p.18) liquid, especially an oil (↑), gives off (p.41) enough vapour to produce a small flame, but not to catch fire, when touched by a small flame or hot object.

ignition point the lowest temperature at which a volatile (p.18) liquid, especially an oil, will catch fire, or burst into flames, when touched by a flame or hot object. **ignite** (*v*).

explosive (*adj*) describes any substance that ignites (p.32) so quickly that it causes an explosion (p.58).

essential oil

COOCH₃
OH

oil of wintergreen
methyl
2-hydroxybenzoate

[C] asymmetric
carbon atom
**hexose, aldose
monosaccharide**

[C] asymmetric
carbon atom
**hexose, ketose
monosaccharide**

**a hexose
an aldose
a monosaccharide**

carbohydrate (*n*) an organic (p.55) substance with the general formula (p.181) $C_x(H_2O)_y$ and a complex molecular structure (p.83). The carbohydrates are divided into two groups, sugars (↓) and polysaccharides (p.207).

sugar (*n*) a colourless, crystalline, water-soluble solid with a sweet taste. Sugars are classified as monosaccharides (↓), disaccharides (p.206), trisaccharides, tetrasaccharides, etc.

monosaccharide (*n*) a sugar (↑) with the formula $C_n(H_2O)_n$, with the most common members of the group having n = 5 for a pentose (↓) or n = 6 for a hexose (↓). The two most common monosaccharides are glucose and fructose (p.206), they are both hexoses (↓). The monosaccharides exhibit stereoisomerism, the isomers are discussed under hexose (↓); monosaccharides are not hydrolyzed to simpler sugars.

pentose (*n*) a monosaccharide (↑) with a molecular formula (p.181) of $C_5H_{10}O_5$. The pentoses can be divided into aldoses (↓) and ketoses (p.206). The most common is ribose.

hexose (*n*) a monosaccharide (↑) with a molecular formula (p.181) of $C_6H_{12}O_6$. The hexoses can be divided into aldoses (↓) and ketoses (p.206). The carbon chains for the two kinds of molecule are shown in the diagram opposite in structural formulae. An aldose (↓) has an aldehyde (p.175) group and a ketose (p.206) has the carbonyl group (p.185) of a ketone (p.176). The hydrogen atoms and hydroxyl groups can have different spatial (p.211) directions, producing stereo-isomers (p.183). Asymmetric carbon atoms (p.183) produce optical isomerism (p.183).

aldose (*n*) a monosaccharide (↑) with an aldehyde group (p.175); it is a reducing sugar (p.206). In a hexose, which is also an aldose, called an **aldohexose**, there are four asymmetric carbon atoms (p.183), none of which is combined with the same organic (p.55) group, hence each asymmetric atom will have two enantiomorphs (p.183). This produces 16 possible isomers; only three occur abundantly (p.231) in nature. Aldoses exhibit (p.221) the reactions of alde-hydes and alcohols (p.175).

ketose (*n*) a monosaccharide (p.205) with a carbonyl group (p.185) giving the reactions of a ketone (p.176). A ketose is also a reducing sugar (↓). In a hexose which is also a ketose, called a **ketohexose**, there are three asymmetric carbon atoms (p.183); each asymmetric atom will have two enantiomorphs (p.183). This produces 8 possible isomers. Ketoses exhibit (p.221) the reactions of ketones and alcohols (p.175).

reducing sugar a sugar (p.205) which reduces Fehling's solution (p.196) and Benedict's solution (p.196). All monosaccharides are reducing sugars, but only some disaccharides.

glucose (*n*) an aldohexose. It is less sweet than sucrose (↓) and very soluble in water; the naturally occurring form is optically active (p.19), rotating the plane of polarized light to the right, and so is also known as 'dextrose'.

fructose (*n*) a ketohexose. It is the sweetest of all sugars and very soluble in water; the naturally occurring form is highly optically active (p.19), rotating the plane of polarized light to the left.

disaccharide (*n*) a sugar (p.205) which consists of two monosaccharides (p.205) chemically combined. On hydrolysis (p.66) one molecule of a disaccharide produces two molecules of monosaccharides; the monosaccharide molecules can be the same, or different. The most common disaccharides have a molecular formula (p.181) of $C_{12}H_{22}O_{11}$.

sucrose (*n*) a disaccharide (↑); on hydrolysis (p.66) it produces equal proportions of glucose (↑) and fructose (↑). A molecule of sucrose consists of a molecule of glucose and a molecule of fructose combined by a condensation (p.191) reaction. Sucrose is a non-reducing sugar (↓).

non-reducing sugar a sugar (p.205) which does not reduce Fehling's solution (p.196) or Benedict's solution (p.196).

maltose (*n*) a disaccharide (↑); on hydrolysis it produces glucose (↑). A molecule of maltose consists of two molecules of glucose combined by a condensation reaction (p.191). Maltose is a reducing sugar.

maltose
disaccharide
(two hexoses combined)

$$CH_2 = CHCl$$
$$+$$
$$CH_2 = CHCl$$
$$+$$
$$CH_2 = CHCl$$
$$+ ...$$

$$\downarrow$$

$$\begin{array}{cc} Cl & Cl \\ | & | \\ CH_2-CH-CH_2-CH.. \end{array}$$

a polymer with **polymer**
900 to 1300
molecules of the
monomer combined

monomer chloroethene

polysaccharide (*n*) a carbohydrate (p.205) that is not sweet, is insoluble in water and is non-crystalline; most polysaccharides are colloidal (p.98) in nature. Polysaccharides formed by condensation reactions (p.191) from hexoses (p.205) have a general formula of $(C_6H_{10}O_5)_n$, where n is a very large number. Starch ($\downarrow$) and cellulose ($\downarrow$) are the most common polysaccharides.

starch (*n*) a polysaccharide ($\uparrow$), which on hydrolysis produces glucose ($\uparrow$). One starch molecule consists of 4000 to 30 000 glucose molecules chemically combined.

cellulose (*n*) a polysaccharide ($\uparrow$), which on hydrolysis produces glucose ($\uparrow$). One cellulose molecule consists of about 3000 glucose molecules.

polymerization (*n*) a chemical process in which molecules of the same compound combine together to form a molecule of high relative molecular mass (p.114). The formation of starch from glucose is a form of polymerization. There are two kinds of polymerization, addition polymerization (p.208) and condensation polymerization (p.208). **polymerize** (*v*), **polymer** (*n*), **polymeric** (*adj*).

monomer (*n*) a molecule or substance which can be polymerized; it has a molecule of low relative molecular mass (p.114), e.g. chloroethene (vinyl chloride) has a molecular formula (p.181) C_2H_3Cl; it is the monomer of polyvinyl chloride (P.V.C.) which contains between 900 to 1300 molecules of the monomer chemically combined.

dimer (*n*) a molecule or compound, formed by the chemical combination of two simpler molecules, e.g nitrogen dioxide, NO_2, forms a dimer, dinitrogen tetroxide, N_2O_4. **dimerization** (*n*), **dimeric** (*adj*), **dimerize** (*v*).

polymer (*n*) a material with a molecule of high relative molecular mass (p.114) formed by polymerization ($\uparrow$). For example, using the monomer ($\uparrow$) chloroethene, a polymer is produced with a relative molecular mass of 50 000 – 80 000. The constitutions of the molecules of a polymer vary, it is described as being between a range of relative molecular masses.

$$CH_2 = CH_2 + CH_2 = CH_2 + CH_2 = CH_2 + \cdots$$

ethene (monomer)

↓

$$-CH_2-CH_2-CH_2-CH_2-CH_2-CH_2-$$

addition polymerization   polythene (polymer)

addition polymerization polymerization (p.207) in which monomers (p.207) combine together by an addition reaction, so that the polymer has the same empirical formula (p.181) as the monomer. For example, polyvinyl chloride is a polymer (p.207) with an empirical formula of $(C_2H_3Cl)_n$, where n is between 900 and 1300; the empirical formula of the monomer is C_2H_3Cl. Addition polymerization takes place with alkenes (p.173) and their derivatives (p.200). Chloroethene (C_2H_3Cl) is the monomer (p.207) of polyvinyl chloride.

condensation polymerization polymerization (p.207) in which monomers (p.207) combine together in a condensation reaction (p.191). For example, aminoethanoic acid (glycine) undergoes (p.213) condensation polymerization. Most condensation polymerizations are a kind of copolymerization (↓).

copolymerization
(also a condensation polymerization)

methyl terephthalate   ethane-1, 2-diol

$$CH_3OOC-\bigcirc-COOCH_3 + HO\,CH_2\,CH_2OH$$

↓ $-CH_3OH$

$$-OC-\bigcirc-COOCH_2\,CH_2OOC-\bigcirc-OC-$$

terylene

copolymerization (*n*) polymerization (p.207) using two, or more, monomers (p.207); e.g. the polymer terylene starts with two monomers: dimethyl benzene-1,4-dicarboxylate (methyl terephthalate) and ethane-1,2-diol. Condensation polymerization (↑) takes place with the elimination (p.189) of methanol.

polythene (*n*) a polymer (p.207) which has ethene (p.174) as a monomer (↑). Molecules of ethene in the presence of a catalyst undergo (p.213) addition polymerization (↑). Polythene is inert (p.19), a good electrical insulator, and can be moulded (p.210).

peptide (*n*) a polymer (p.207) formed by the condensation copolymerization (↑) of several amino acids (p.178) joined by a peptide bond (↓). On hydrolysis it produces amino acids.

peptide bond a condensation reaction (p.191) between an amino group (p.186) and a carboxyl group (p.186) with the elimination (p.189) of water. In hydrolysis the peptide bond is broken to form the original (p.220) amino acids.

protein (*n*) a polymer (p.207) formed from peptides (↑) by condensation polymerization (↑). A protein contains 50 or more amino acids (p.178) joined by peptide bonds (↑). On hydrolysis a protein is first decomposed to peptides and then to amino acids.

rubber (*n*) a naturally occurring (p.154) elastic material which is a polymerized (p.207) hydrocarbon (p.172). Destructive distillation (p.203) of natural rubber produces 2-methylbuta-1,3-diene. Synthetic rubber is generally manufactured (p.157) by the copolymerization of phenylethene (styrene) and buta-1,3-diene (butadiene). *See diagram.*

rubber

$$CH_2 = C - CH = CH_2$$
$$|$$
$$CH_3$$

2-methyl buta-1, 3-diene
monomer of natural rubber

$$CH_2 = CH \quad + \quad CH_2 = CH - CH = CH_2$$

phenyl
ethene

buta-1, 3-diene

$$-CH_2 - CH - CH_2 - CH = CH - CH_2 -$$

synthetic rubber

glasses (*n.pl.*) amorphous (p.15) solids composed of silicon dioxide (SiO_2) with the silicon and oxygen atoms forming a tetrahedral (p.83) structure. Cations of various metals form bonds with the atoms in the tetrahedral structure. Like polymers (p.207) glasses have molecules of very high relative molecular mass (p.114). Soda glass, the normal soft glass, is manufactured (p.157) by fusing together a mixture of sand, limestone and sodium carbonate. Glasses have no fixed melting point (p.12).

plastic[2] (*n*) a material manufactured (p.157) by
· the polymerization (p.207) of organic (p.55)
substances. It is usually a hard material, or a
threadlike material used for making cloth, and
does not appear to be plastic (p.14) in the
usual meaning of the word. A plastic material
has its shape changed, either when hot or
cold, and then retains (p.215) this shape. A
manufactured plastic has the property of
plasticity at some stage of its manufacture.
There are two types of plastics, thermoplastics
and thermosetting plastics.

thermoplastic (*n*) a kind of plastic which exhibits
(p.221) plasticity when hot. The material is
heated and pressed into shape in a mould (↓).
On cooling it retains the shape of the mould.
On heating again, its shape can be changed
by pressure, e.g. polyvinyl chloride (P.V.C.)
and polythene (p.208). These materials are
soluble in organic solvents.

thermosetting plastic a kind of plastic which
exhibits plasticity (p.14) when first heated; by
heating, the structure of the polymer is
changed and bonds are formed between
molecules to produce a three-dimensional
structure, making the plastic very strong when
it has cooled. When the plastic is heated
again, its shape cannot be altered, i.e. it has
lost its plasticity, e.g. urea-methanal resins and
Bakelite.

mould (*n*) a vessel into which a hot liquid
material is poured to become cold. When cold
it solidifies (p.10) in the shape of the mould.
Powders can be used instead of liquids and
then pressure and heat are applied (p.232) to
turn the powder into a solid of the same shape
as the mould. This method is used with
plastics (↑). **mould** (*v*), **moulding** (*n*).

plasticizer (*n*) a substance which is added to a
polymer (p.207) to keep it plastic, e.g. a
plasticizer is added to polyvinyl chloride to
alter its properties. With a small quantity of
plasticizer, a strong solid is produced; with a
large quantity of plasticizer, an elastic solid is
produced.

powdered plastic put
in mould

mould heated
pressure applied

mould
cooled

moulded
plastic

moulding plastic

thermoplastic

**thermosetting
plastic**

heated
again

shape can
be changed

shape
unchanged

position of end

position

spatial

spatial arrangement
of cross pieces

structure
of a bridge

position (*n*) the position of an object is its place in space in relation (p.232) to other objects. For example, (a) the position of a table in a room is its place in relation to a door, a window, or chairs in the room; (b) the position of an element in the electrochemical series (p.130) shows the relation of that element to the other elements in the series, i.e. whether it has a greater or lesser electrode potential. **position** (*v*).

spatial (*adj*) describes an arrangement or a direction or an extent (p.213) in space, e.g. the spatial properties of covalent bonds (p.136) describe the direction and arrangement of the bonds in space. **space** (*n*), **space** (*v*).

limit[1] (*n*) the value of a quantity (p.81) beyond which it is generally not possible to go. If a limit can be exceeded, *see diagram*, then different circumstances or different physical laws act on the quantity. A limit can be the largest or smallest possible value, e.g. the limit of solubility of a crystalline substance is the greatest amount that dissolves in boiling water.

structure[2] (*n*) the arrangement in space of the connected (p.24) parts of a whole object. For example, (a) the structure of a bridge is the arrangement in space of its different parts; (b) the structure of a molecule of ethanol (p.175) shows the arrangement in space of the atoms in the molecule. **structural** (*adj*).

structural (*adj*) describes anything to do with structure, e.g. structural isomerism (p.182) is a kind of isomerism which depends on molecules having different structures.

construct (*v*) to make a structure, e.g. to construct a model (p.223) is to make a model with a particular structure. **construction** (*n*).

slit (*n*) a long, very narrow hole in a surface.

limit

within the limit	beyond the limit
at the limit	
exceeding the limit	

0   10   20   30   40
km/hr

slit in a surface

slit

system (*n*) (1) a fixed way of carrying out (p.157) a process (p.157), e.g. the system of naming chemical substances. (2) a set of objects which obey physical laws, have an effect on each other and form a whole unit (↓), e.g. the substances in an equilibrium mixture (p.150) form a system. **systematic** (*adj*).

unit (*n*) (1) a value of a quantity which is accepted as a standard (p.229) for that quantity, e.g. the kilogram is accepted as the unit of mass. (2) a whole thing, made from different parts, which acts as a whole, e.g. a molecule (p.77) of a substance is made of different parts (the atoms) and it acts as a single object, it forms a unit. (3) each member of a series (p.172) is a unit.

circumstances (*n.pl.*) everything that may or may not have an effect on an object or a substance, together make the circumstances of that object or substance; e.g. the circumstances of a liquid are: the vessel containing it, the temperature, the atmospheric pressure, etc. Contrast *conditions*: *circumstances* that have an effect on a process involving the object or substance are its *conditions*.

local (*adj*) describes anything near an object or substance, e.g. the local conditions or the local circumstances. *Contrast* ambient (p.103) which describes anything surrounding an object.

general (*adj*) describes properties, qualities and nature possessed by all members of a set, e.g. the properties which all acids possess and their general properties. *Contrast* the additional properties of any one acid which are its **particular** properties. **generalize** (*v*), **generalization** (*n*).

special (*adj*) describes properties, qualities and nature which are very different for one member of a set and distinguish (p.224) it from all other members of the set.

universal (*adj*) (1) describes a statement which is always true and has no exceptions (p.230), e.g. a universal law, such as the attraction between unlike electric charges, has no exceptions. (2) describes a class of substances which have a characteristic action without exception, e.g. a universal solvent for organic substances.

use of
 general
 particular
 special

sulphuric acid   hydrochloric acid

acids general properties

special properties

nitric acid   particular properties

exist (v) to be, of objects, materials and substances, i.e. to be perceptible (p.42) to sight, hearing, feeling, tasting or smelling. For example, carbon exists in two crystalline forms. *See occur (p.154)*. **existence** (n).

extend (v) (1) to take up space between two points or to cover an area, e.g. the coal seam (p.154) extends from a depth of 200 metres to a depth of 250 metres. (2) to take up time between two points in time, e.g. the whole process of separating the mixture by fractional distillation extended from 2.00 p.m. to 4.30 p.m. **extent** (n).

extent (n) (1) the space, time or activity between limits, e.g. the extent of the decomposition of phosphorus pentachloride into phosphorus trichloride and chlorine depends on the temperature and pressure (the temperature and pressure make the limits). (2) the space between limits over which something extends (↑). **extensive** (adj), **extensively** (adv).

accompany (v) to take place (p.63) or to exist (↑) at the same time, e.g. the reaction between metals and concentrated acids is usually accompanied by the evolution (p.40) of heat. **accompaniment** (n).

limit[2] (v) to make a limit (p.211) for time or activity, e.g. (a) refluxing (p.201) the mixture was limited to half an hour; (b) pressure limits the formation of ammonia from nitrogen and hydrogen.

effort (n) the use of force or energy, e.g. an effort is needed to keep the reactants in a chemical process under pressure.

undergo (v) to take part in an action or a chemical reaction because of an outside agent (p.63), e.g. iron undergoes rusting in damp air (the damp air is an outside agent causing the rusting).

overcome (v) to make an action take place when there is resistance (↓) to the action, e.g. sufficient energy has to be supplied (p.154) to reactants to overcome the energy barrier (p.152).

resist (v) to try to prevent an action taking place, e.g. a chemical reaction does not take place until the energy barrier (p.152) is overcome (↑); the energy barrier resists the reaction taking place. **resistance** (n).

extent

the extent of a coal seam

extent

upper limit

lower limit

coal

overcome

man overcomes resistance – case moves

case resists moving

resist

cause (*n*) a substance, a form of energy, or an event which makes a change, a process, or another event take place. For example, (a) oxygen in the air is a cause of iron rusting; (b) a flame is the cause of many explosions of methane in coal mines; (c) the escape of coal gas was the cause of the explosion. **cause** (*v*).

effect (*n*) the change, process or event produced by a cause (↑), e.g. heat causes water to boil, the boiling of the water is the effect of the heat.

effective (*adj*) describes anything that produces an effect (↑). For example, (a) the hydroxyl group in an alcohol (p.175) is the effective part of the molecule; (b) propanone (acetone) is a very effective solvent for organic substances (it produces the effect of dissolving the solute) **effectiveness** (*n*).

efficient (*adj*) describes a device, a piece of apparatus, or a process which produces a result with little or no waste of energy or material. For example, (a) a fractionating column is an efficient piece of apparatus for separating a liquid mixture; (b) the Solvay process (p.169) is the most efficient way of manufacturing sodium carbonate. **efficiency** (*n*).

facilitate (*v*) to make an action or a process take place more easily, e.g. a catalyst facilitates the chemical combination of nitrogen and hydrogen in the Haber process (p.170). **facility** (*n*).

favour (*v*) to provide the conditions (p.103) so that a chemical reaction takes place more easily or more quickly, e.g. high pressure favours the combination of nitrogen and hydrogen to form ammonia; high pressure is the condition, it does not cause the reaction to take place more easily, but the pressure offers less resistance (p.213) to the change. **favourable** (*adj*).

terminate (*v*) to stop a process, or a chemical reaction, before it has had time to finish, e.g. the distillation was terminated when the flask cracked. **termination** (*n*).

duplicate (*v*) to carry out the same action twice, e.g. the experiment was duplicated so that the two sets of results could be compared (p.224). **duplicate** (*n*), **duplicate** (*adj*).

cause

direct cause

effect

one key is the duplicate of the other

duplicate

removing an
electric light bulb

exchanging two
electric light bulbs

exchange

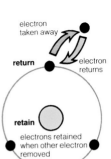

electron
taken away

return electron
returns

retain

electrons retained
when other electron
removed

remove (*v*) to take an object, a material, or a substance from a place, e.g. (a) to remove a residue (p.31) from a filter paper; (b) to remove impurities from a metal. **removal** (*n*), **removable** (*adj*).

substitute (*v*) to put one object in the place of another object, when the two objects are not similar and do not have the same properties, e.g. to substitute a chlorine atom for a hydrogen atom in an organic compound (the two atoms are not similar and do not have the same properties). **substitution** (*n*).

exchange (*v*) to put one object in the place of another object when the two objects are not similar but do have the same use or property, e.g. when a solution of ammonium sulphate passes through soil, calcium sulphate solution passes out of the soil; the ammonium ions have been exchanged for calcium ions (the two ions are not similar, but they have the same properties of being cations (p.125). **exchange** (*n*).

interchange (*v*) to put one object in the place of another object when the two objects are identical, e.g. when water is in equilibrium (p.150) with its vapour in a cloud vessel, the molecules in the water interchange with the molecules in the vapour. **interchange** (*n*).

retain (*v*) to continue to possess an object, energy, substance or property when conditions try to remove (↑) the object, energy, substance or property. For example, (a) after distillation, the flask always retains some liquid; (b) after purification of pig iron to wrought iron (p.163) the iron still retains some impurities. **retention** (*n*).

revert (*v*) to go back to an original (p.220) state, e.g. on heating rhombic sulphur slowly it changes to monoclinic sulphur at 96.5°C. On cooling, monoclinic sulphur slowly reverts to rhombic sulphur.

return (*v*) to go back to an original (p.220) place, e.g. if an electron is removed from an atom, a positive ion is formed; if an electron returns to the ion, the atom is formed.

exit (*n*) a place through which something goes out, e.g. in a fractionating column there are several exits, one for each fraction. **exit** (*v*).

prevent (*n*) to cause an action or a process not to take place or not to happen. An agent is needed as the cause. The presence (↓) of the agent, or the action of the agent, is the cause. The agent can be an object, substance, condition or a person. **prevention** (*n*).

interfere (*v*) to cause a process to become slower, or to stop, or make it difficult to observe (p.42). The process is usually one that is wanted, e.g. (a) the presence of propane-1,2,3-triol (glycerol) interferes with the decomposition of hydrogen peroxide, the process is slowed down; (b) the presence of a sodium salt interferes with the flame test for other metals (makes the test difficult to observe). **interference** (*n*).

counteract (*v*) to act against a process so that the process is slowed down or stopped, or made to go in the reverse (↓) direction. The process is usually, but not necessarily, one that is not wanted, e.g. substances are added to rubber to prevent atmospheric oxygen destroying its properties, these substances counteract the effect of atmospheric oxygen. **counteraction** (*n*).

reverse (*v*) to make a process go in the opposite direction, e.g. heating a liquid causes it to vaporize (p.11), cooling the vapour reverses the process and the liquid is condensed. **reverse** (*n*), **reverse** (*adj*).

tend (*v*) to have a possible action or behaviour, with the action taking place slowly, or not taking place if conditions (p.103) are not suitable; e.g. a solution of sodium hydroxide tends to absorb carbon dioxide from the air, this can be counteracted (↑) by using a tight-fitting rubber bung (p.24). **tendency** (*n*).

trend (*n*) the general direction of change in a set of related facts, e.g. (a) in homologous (p.172) series the trend is for a decrease in activity with an increase in the number of carbon atoms (the general direction of related change); (b) in the set of alkali metals (p.117) the trend is for increasing ease of ionization with increasing atomic number.

colour of sodium
flame interferes
with test for copper

present[1] (*v*) to produce a thought in an observer, e.g. (a) making volatile (p.18) compounds react when heated presents a difficulty to an observer which he overcomes (p.213) by refluxing (p.201) the compounds; (b) finding a use for waste products presents a problem to a manufacturer.

present[2] (*adj*) describes an object or substance which is in a particular, named place. For example, (a) the iodine test shows whether starch is present in food; (b) the presence of chloride ions (p.123) is tested by silver nitrate solution. **presence** (*n*).

recur (*v*) to occur (p.63) time after time, usually at a definite interval (p.220) of time, e.g. conditions for the pollution of the atmosphere recur every year when the temperature is suitable for the formation of oxides of nitrogen.

flow (*v*) to move, of a liquid or gas, along a pipe or over a surface. Electric current and heat flow along a conductor. **flow** (*n*).

obstruct (*v*) to prevent (↑) the flow of a liquid or gas, or the flow of a stream of particles (p.110), e.g. a deposit of calcium carbonate in a water-pipe obstructs the flow of water. **obstruction** (*n*).

stoppage (*n*) the state of a flow being stopped, e.g. a stoppage in the delivery tube connecting a distillation flask to a condenser means that the flow of vapour has stopped and an explosion is likely.

capture (*v*) to attract and to hold an object by the use of force. To catch and to keep an escaping object. For example, (a) an atomic nucleus (p.110) captures a neutron (p.110) during bombardment (p.143) by neutrons (the neutron is held in the nucleus by nuclear forces); (b) a steam trap (p.29) captures any water escaping with the steam. **capture** (*n*).

contact (*n*) the state of two, or more, objects being in touch with each other. For example, (a) an electrical contact is formed when two conductors (p.122) touch each other; (b) when sulphur dioxide and oxygen are both in contact with a platinum catalyst, the two gases combine (the two gases touch each other and the catalyst). **contact** (*n*).

pipe water flowing

obstruct

flow

obstruction

flow
obstructed

point of
contact

two metals
in contact

contact

overlap (*v*) to cover part of a flat object by putting another flat object above it, e.g. the pieces of wood on the side of a boat overlap each other. **overlap** (*n*).

rotate (*v*) to turn, of an object, if the object turns about an axis going through the body, e.g. the Earth rotates about its axis. **rotation** (*n*).

vary (*v*) to change in detail (p.226) only, of a quantity (p.81), a quality (p.15), or a shape, e.g. (a) atmospheric pressure varies from day to day; (b) the electric current passing through an electrolyte (p.122) varies with the voltage (p.126) applied to the electrodes (the process of electric current being conducted by the electrolyte remains the same, only the detail of the value of the current varies). **variation** (*n*), **variable** (*adj*), **variable** (*n*), **varying** (*adj*), **varied** (*adj*).

variation (*n*) (1) the amount of change which takes place if a quantity or quality varies (↑); e.g. the variation in atmospheric temperature from the highest to the lowest temperature of the day, (2) the action of varying (↑).

variable[1] (*n*) a quantity, such as temperature, pressure, humidity, concentration, etc. which can change in value or can be changed in value, e.g. the vapour pressure (p.103) of a liquid is a variable because it varies (↑) with temperature.

variable[2] (*adj*) describes a quantity, quality, or shape which varies (↑) or can be varied, e.g. atmospheric pressure is variable as it changes from day to day.

maximum (*n*) (maxima *n.pl.*) the greatest possible, or the greatest recorded (p.39), value of a variable (↑), e.g. (a) 41°C was the maximum recorded for atmospheric temperature in the country; (b) the maximum pressure the chamber will withstand is 200 atmospheres.

minimum (*n*) (minima *n.pl.*) the least possible, or the lowest recorded, value of a variable (↑), e.g. 630 mm of mercury was the minimum recorded for atmospheric pressure in the country; (b) the minimum pressure recorded in the vacuum distillation was 50 mm of mercury.

two metals overlap

overlap

Earth rotates

rotate

rotation   axis of rotation

fractionating column

alter

one alteration

one alteration

column modified to collect two distillates

modify

alter (*v*) to make one variation (↑) in a quantity, shape, or condition, e.g. (a) to alter the pressure of a gas from 1 atmosphere to 2 atmospheres; (b) to alter the condition in a reversible reaction by the addition of a catalyst (change of conditions). **alteration** (*n*).

modify (*v*) to alter (↑) a process, or an object, to make it more suitable for a particular purpose, e.g. a fractionating column (p.201) can be modified to produce two fractions (p.202) instead of one (both a process and an object are altered). If a process or an object is modified, its purpose is not changed. **modification** (*n*).

decrease (*v*) to become, or to cause to become, less, of a quantity (p.81) or number. For example, (a) the rate of reaction (p.149) between ethanol and ethanoic acid is decreased by lowering the temperature of the reactants (p.62); (b) the volume of a gas decreases as the pressure on the gas increases. **decrease** (*n*).

reduce (*v*) to cause a quantity (p.81), or number to become less, e.g. the concentration of the solution was reduced from 2.0 M to 1.0 M by adding a suitable volume of solvent. To contrast *decrease* (↑) and *reduce*: lowering the temperature by 80°C *decreases* the volume of the gas by 160 cm³. Lowering the temperature of the gas by 80°C *reduces* the volume to 5.30 dm³. *Decreasing* describes a continuous change; *reducing* describes a change from one state, or degree (p.227), to another. **reduction** (*n*).

accumulate (*v*) to increase a quantity (p.81) or an amount by addition over a period of time, e.g. (a) during electrolysis (p.122) the deposit on an electrode accumulates as the process goes on; (b) as a suspension settles, the sediment accumulates at the bottom of the vessel. **accumulation** (*n*).

accelerate (*v*) to increase speed with time, e.g. some catalysts accelerate the rate of reaction more than others do, i.e. they increase the rate of reaction.

event (*n*) a change that takes place (p.63) at a particular time and usually at a particular place, e.g. (a) the appearance of a precipitate is an event; (b) an explosion is an event. **eventual** (*adj*), **eventually** (*adv*).

interval (*n*) the distance between two points or the length of time between two events.

duration (*n*) the length of time in which a process takes place, e.g. (a) the duration of a process of distillation; (b) the duration of a chemical reaction. To contrast *duration* and *extent* (p.213): The *duration* of the distillation is one hour, the *extent* of the distillation is from the time of starting to heat the flask until the last drop of distillate is collected, i.e. *duration* gives the time for the process, while *extent* gives the limits of the process. **durable** (*adj*), **endure** (*v*).

precede (*v*) to come immediately before an event in time, or before a unit (p.212) in a series (p.172), e.g. number 7 precedes number 8 in the series 5, 6, 7, 8, 9, 10. Numbers 5, 6, 7, are the preceding terms for number 8, while 7 is the preceding term. **preceding** (*adj*).

previous (*adj*) describes events which come before a named event, or a named time, or the present time, e.g. in a previous experiment, which took place last year, the results were different. To contrast *previous* with *preceding*: the *preceding* (↑) experiment is one immediately before the present experiment; a *previous* experiment is one which took place some time ago. **previously** (*adj*).

preliminary (*adj*) describes the first stage of a process, e.g. preliminary tests in analysis indicate (p.38) the subsequent (↓) stages that have to be carried out.

original (*adj*) describes the very first event, action or process, e.g. Becquerel carried out the original experiments on radioactivity.

subsequent (*adj*) describes events which come after a named event, or a named time, or the present time, e.g. the results of the present experiment are not sufficient for a conclusion, subsequent experiments may give a better result. **subsequently** (*adv*).

interval

event

(circuit)

circuit to bulbs

the bulbs light simultaneously with the closing of the switch

(circuit)

simultaneous

simultaneous (*adj*) describes two or more events which take place at the same time, e.g. the closing of a switch in an electrical circuit and the flow of electric current in the circuit are simultaneous. **simultaneously** (*adv*).

immediate (*adj*) describes one event which is next in time to another event without anything happening between the two events, e.g. precipitation is immediate when silver nitrate solution is added to the solution of a chloride. **immediately** (*adv*).

order (*n*) the arrangement of parts of a whole, or of members of a series (p.172), so that a pattern (p.93) is formed; or the arrangement follows particular rules; or the arrangement occurs (p.63) in time. For example, (a) the molecules in a giant structure are arranged in a particular pattern which has a definite order; (b) the stages of distillation follow each other in order; (c) words can be arranged in an alphabetical order. **ordered** (*adj*).

method (*n*) a particular way of carrying out a process, e.g. (a) the method of separating a liquid mixture by fractional distillation; (b) copper (II) sulphate can be prepared by digesting copper (II) oxide in sulphuric acid or by digesting copper (II) carbonate in sulphuric acid; these are two different methods.

control (*v*) to start or stop a process, to increase or decrease the rate of a reaction; to vary a quantity for a purpose; e.g. to control the manufacture of ammonia by using suitable conditions of temperature and pressure and adding suitable amounts of reactants (p.62). **control** (*n*).

progress (*v*) of a process, to go successfully from one stage (p.159) to the next stage in the correct order (↑), e.g. as the distillation progressed, the distillate (p.201) accumulated (p.219) in the receiver. **progress** (*n*).

exhibit (*v*) to allow a property (p.9) to be observed when it is not always open to observation (p.42). For example sodium carbonate crystals exhibit efflorescence (p.67), i.e. they can be observed to effloresce only under particular conditions.

statement (*n*) a description of related facts in language or in symbols, e.g. a statement of Boyle's law in language or in symbols. The related facts are the mass of the gas, the pressure, its volume and its temperature. **state** (*v*).

information (*n*) the knowledge given by a statement (↑), e.g. information on the change in volume of a gas under particular conditions can be obtained from a statement of the gas laws (p.109). **inform** (*v*).

determine (*v*) (1) to find a value of a quantity if more than one measurement has to be made and the final value obtained by calculation, or to find a very accurate value of a quantity by careful measurement, e.g. to determine the boiling point of a substance using a very accurate thermometer and special apparatus. **determination** (*n*). (2) if one variable (p.218) depends on another variable, then the second variable determines the value of the first variable, e.g. the volume of a gas depends on its pressure, so its pressure determines its volume.

deduce (*v*) to come to a conclusion (p.43) using known facts, e.g. fact 1: substance X neutralizes an alkali; fact 2: acids neutralize alkalis; deduce that substance X is an acid. **deduction** (*n*).

verify (*v*) to carry out an experiment and to record results using materials and substances, when similar experiments have been carried out previously (p.220), to show that the previous results, and the deductions (↑) from the results, were true. For example, to carry out an experiment to verify the statement (↑) of Boyle's law. **verification** (*n*).

assume (*v*) to take a statement (↑) or a fact, without verification (↑) to be true, e.g. in a calculation on relative vapour density (p.12) we assume the gas laws. **assumption** (*n*).

clarify (*v*) to make a statement (↑) clear enough to understand it, e.g. a statement about the direction of the covalent bonds of a carbon atom is clarified by making a drawing of the atom and bonds. **clarification** (*n*).

deduce

fact 1

substance x

alkali
(neutralized)

acid neutralizes
alkali   fact 2

acid

fact 1 + fact 2
↓
deduce
↓
x is an acid

an adequate
model of ethane

model (*n*) a device (p.23) which represents (↓) an object and allows a person to think about the object when the object itself cannot be observed (p.42). A **scale model** is an exact (p.79) representation with each part a known fraction, in size, of the object, e.g. a scale model of a house. An **adequate model** has sufficient (p.231) detail for a particular purpose, e.g. an adequate model of an atom describes the nucleus (p.110) and extra-nuclear electrons (p.110). An **analogue model** uses only similar characteristics, e.g. a permeable membrane is represented as a sieve with holes.

represent (*v*) to make a drawing which is like an object and allows a person to remember or to think about the object, e.g. the drawings of apparatus on page 23 represent each piece of apparatus; they are not the same as pictures of the apparatus, but the representation allows a person to imagine the actual piece of apparatus. **representation** (*n*).

apparent (*adj*) describes anything that appears to be correct when observed (p.42) by the senses, but can be shown by experiment, or is known, not to be correct, e.g. the apparent r.m.m. of ethanoic acid, when dissolved in some solvents, is twice the actual value, i.e. the measurement does not give a correct value.

recapitulate (*v*) to state (↑) facts a second time, usually in a different way, in order to clarify (↑) the information.

refer (*v*) to bring facts, statements, or information to a person's attention, e.g. when discussing electrovalency it is useful to refer to the atomic structure of elements, i.e. to direct a person's attention to the information on atomic structure.

random (*n*) describes an event, a process or a state which exhibits (p.221) no order (p.221), e.g. it is not possible to predict (p.85) the path that the molecules of a gas will follow.

probability (*n*) the mathematical fraction which describes chance, e.g. when a die is thrown, there is an equal chance that any one of the six faces will appear; the probability of any one particular face appearing is 1/6.

die

compare (*v*) to list and discuss the similar
(p.233) properties and characteristics of two
objects, or the similar characteristics of two
processes, e.g. to *compare* sodium *to*
potassium. If both similar and different
properties or characteristics are discussed
then one thing is *compared with* another.
comparison (*n*), **comparable** (*adj*).

contrast[1] (*v*) to list and discuss the different
properties and characteristics of two objects
or the different characteristics of two
processes. **contrast** (*n*), **contrasting** (*adj*).

contrast[2] (*n*) (1) a list of the differences between
two objects or processes. (2) the existence of
differences between two objects or processes.

distinguish (*v*) to recognize differences in detail
between two substances, two objects or two
kinds of radiation (p.138) when the
substances, objects or radiations are similar in
most properties or characteristics. For
example, (a) the properties of ethanol and
propanol are very similar but they can be
distinguished by their boiling points; (b) a
person can *distinguish between* kerosene and
petrol; he can *distinguish* kerosene *from*
petrol. **distinction** (*n*), **distinguishable** (*adj*),
distinct (*adj*).

distinction (*n*) a difference between two
substances, two objects or two kinds of
radiation (p.138) which allows the substances,
objects or radiations to be distinguished (↑),
e.g. distinctions can be clear or slight; a
distinction can be made or can be drawn.

distinctive (*adj*) describes a property or
characteristic which allows a substance, an
object or a kind of radiation (p.138) to be
distinguished (↑) easily from all others, e.g.
chlorine has a distinctive odour (p.15) which
allows it to be easily recognized.

distinct (*adj*) describes an object or a kind of
radiation (p.138) which possesses at least one
distinctive (↑) property or characteristic, e.g.
(a) copper is distinct among all metals
because of its colour; (b) chlorine is distinct
amongst gases because of its odour.

compare A to B
A, B similar shape A, B same colour

compare A with B
A, B similar shape A, B same colour A is big B is small

contrast A with B
A is big B is small

C, D same shape
   same colour
C, D distinguished by
   small black circle
   (small detail)

distinguish

electroscope

rays
detected

α-ray

radioactive
source

detect

differentiate (*v*) to give reasons why a
substance, an object or a kind of radiation is
different from other similar objects or other
similar radiations. For example, Boyle
differentiated between elements and
compounds by saying elements could not be
decomposed whereas compounds could be
decomposed. **differentiation** (*n*).

detect (*v*) to discover the presence of a substance, an object, or a kind of radiation by the
use of particular tests, skilfully carried out, e.g.
(a) to detect the presence of impurities in a
specimen of silver; (b) to detect alpha radiation
from a radioactive source. **detection** (*n*).

identify (*v*) to name a specimen of a substance,
a process, or a radiation after comparing it to a
known substance, process or radiation, and
observing that the properties and
characteristics are identical (p.233). For
example, a colourless liquid is volatile and
flammable, it has a boiling point of 78°C,
reacts with ethanoic acid to form an ester, is
oxidized to ethanoic acid and reacts with
phosphorus pentachloride. These are the
properties of ethanol, so the liquid is identified
as ethanol. **identification** (*n*), **identify** (*n*),
identifiable (*adj*).

identification (*n*) the process of identifying (↑) a
substance, process, or radiation.

identity (*n*) the name of a substance, process or
radiation or a description of the properties and
characteristics which allow it to be used for
identification subsequently, e.g. when
Becquerel discovered radiation from
radioactive sources, the true identity of the
radiation was not immediately established (↓).

establish (*v*) (1) to make the identity (↑) of a
newly discovered particle, substance, process
or radiation, accepted, e.g. Chadwick
established the identity of the neutron by his
experimental results. (2) to make a theory or
hypothesis accepted, e.g. Avogadro's
hypothesis was not established until its
usefulness in determining relative molecular
masses became known. **establishment** (*n*).

detail (*n*) a small part of a structure, object, or process; or a fact, which is not important, in a set of facts. For example, (a) in describing the structure of a molecule of ammonia, the angles between the bonds are a detail; (b) in describing the preparation of copper (II) sulphate from copper (II) oxide and sulphuric acid, an experimental detail is that the oxide should be added in small amounts (a part of a process, or a fact in a set of descriptive facts).

detailed (*adj*) describes a statement (p.222) which gives as many details as possible.

essential (*adj*) describes any part of a whole without which the whole loses its identity (p.225). An essential property or characteristic is one that must be possessed for the purpose of identification, e.g. (a) the amino group (p.186) is an essential part of the molecule of an amine (p.178); without it the substance is no longer an amine; (b) the ability to produce hydrogen ions in solution is an essential property of an acid; if a substance does not exhibit (p.221) that property it cannot be classified (p.120) as an acid. **essentially** (*adv*).

major (*adj*) describes any part, property, characteristic or fact which is important or is the most important, e.g. (a) the major use of chlorine is in the manufacture of plastics and synthetic rubber (fact); (b) the major source of bromine is sea water. **majority** (*n*).

minor (*adj*) describes any detail, property, characteristic or fact which is not important or is of lesser importance. For example, (a) a minor use of lead is in lead tetraethyl, added to petrol; (b) a minor detail in the process of distillation is controlling (p.221) the flow of water through the condenser to obtain the best conditions for condensation. **minority** (*n*).

definite (*adj*) describes a statement, a relation (p.232), a property or a characteristic about which there is no doubt, as previous experimental work has established (p.225) the facts, e.g. aluminium oxide has a definite giant covalent crystalline structure, i.e. the structure has been established (p.225).

detail

detail of water level

detail of scale

measuring cylinder

degree (*n*) (1) a stage on a scale of quality (p.15), e.g. the degrees of a hot object are: lukewarm, warm, hot, boiling, red hot, white hot. A degree describes a simple method of measurement and can be used for quantities (p.81) as well as for qualities. (2) the extent reached by a process, e.g. the degree of ionization measures the extent to which a substance is ionized. This is an accurate (↓) measurement and is usually stated as a percentage or a decimal fraction.

appreciable (*adj*) describes a change in a quantity (p.81) or a quality (p.15) which is big enough or important enough to be considered, e.g. (a) during electrolysis there is an appreciable lowering of the concentration of ions around an electrode; because of this, the electrolyte is stirred; (b) carbon dioxide has an appreciable solubility in cold water, this is important in the carbon cycle (p.61).

accurate (*adj*) describes a measurement made by the best instruments (p.23), or apparatus (p.23), available (p.85), e.g. a pipette (p.26) gives an accurate measurement of volume of a liquid at the temperature marked on the pipette. **accuracy** (*n*).

accuracy (*n*) the quality of being accurate (↑). No measurement can be exact (p.79) as all instruments and scales (p.26) cannot be read to an exact value. In giving a measurement, the degree (↑) of accuracy must be stated (p.222). This is usually done by giving the limits of accuracy, e.g. the accurate value established (p.225) for the Avogadro constant is $(6.022\,52 \pm 0.000\,28) \times 10^{23}\,\text{mol}^{-1}$; the figure $\pm\,0.000\,28$ gives the limits of accuracy, and thus the degree of accuracy of the measurement.

discrepancy (*n*) an appreciable (↑) difference between two measurements or two statements when the measurements should have been the same or the statements should have had the same meaning, e.g. if the results for three titrations of the same solutions are: $21.3\,\text{cm}^3$; $22.8\,\text{cm}^3$; $21.3\,\text{cm}^3$, then there is a discrepancy between the second result and the other two.

accurate

$21.3\,\text{cm}^3$

$21.3\,\text{cm}^3$

$22.8\,\text{cm}^3$ **discrepancy**

define (*v*) to state clearly, in known words, a description of a term which excludes (p.230) similar terms and allows the terms to be discussed (↓) without misunderstanding. Examples of terms are quantities, units of measurement, categories of objects and substances, properties and characteristics. **definition** (*n*).

definition (*n*) the result of defining a term. It can also include giving a name to a new idea, such as the naming of the neutron. Examples are: the definition (a) of an ion using the words: atom, electric charge, positive and negative, electron; (b) of electrochemical equivalent using the words: mass, deposit, coulomb.

discuss (*v*) to examine a statement and to give reasons, in writing or by speaking, for the statement, or against it, e.g. to discuss the statement that fractional distillation is not necessary to separate two liquids if their boiling points are 10°C apart (there are reasons for and against the statement). **discussion** (*n*).

comment (*v*) to explain, in writing or by speaking, why a statement is true or untrue, giving reasons for what is said, e.g. to comment on Boyle's law. **comment** (*n*).

exemplify (*v*) to give examples in order to clarify (p.222) a statement, e.g. the dehydrating properties of concentrated sulphuric acid can be exemplified by its action on sucrose, in which the sugar is converted to carbon. **exemplification** (*n*).

complex (*adj*) describes a structure, process or system formed from parts connected or combined, in an order that can be recognized, to form a whole, e.g. a polymer (p.207) molecule is complex; it consists of small molecules (the parts) combined into a whole, and the structure of the whole molecule can be recognized from the combination of the parts. **complexity** (*n*), **complex** (*n*), **complex** (*v*).

simple (*adj*) describes anything that is not made from parts, is not complex (↑), is easy to understand, e.g. compare a complex ion with a simple ion. **simplicity** (*n*).

complex

complex structure of a molecule

HO

CH₃

OH

parts are 1

2

3

simple

CH₃

a simple molecule

uniform

uniform colour

non-uniform

non-uniform colour

uniform (*adj*) describes a constant (p.106) value of a quantity, or quality, which is spread over space or time, e.g. (a) a uniform colour is a constant colour spread over a surface; (b) a uniform acceleration is a constant acceleration spread over a period of time. To contrast *uniform* and *constant*: there is a *constant* concentration of carbon dioxide in the air (it is 0.03%); there is a *uniform* concentration of carbon dioxide in the atmosphere (the same concentration spread over the surface of the Earth). **uniformity** (*n*).

steady (*adj*) describes a value of a quantity which is kept constant (p.106) although there is a tendency (p.216) for the value to vary, e.g. to maintain a steady vacuum in vacuum distillation (the pressure tends to rise).

standard[1] (*adj*) (1) describes the value of a quantity which is accepted by everyone as a constant (p.106) value, e.g. (a) standard atmospheric pressure is accepted as 760 mm of mercury; (b) standard electrode potentials (p.130) are those defined (↑) under known conditions of concentration and temperature. (2) describes an instrument (p.23) or a piece of apparatus (p.23) having a stated value of a quantity, or an accurate scale for a quantity, e.g. a standard graduated flask with an accurate volume at a given temperature. **standard** (*n*), **standardization** (*n*), **standardize** (*v*).

standard[2] (*n*) an accepted value of a quantity given in a definition, e.g. the standard for length is the metre, defined in S.I. units (p.212).

normal (*adj*) describes conditions, particularly room conditions, which are between the expected limits of variation, e.g. in northern countries, the normal room temperature is accepted to be 18°C; the temperature varies daily above or below this value, but on most days the temperature will not be too different from this value. **normality** (*n*).

abnormal (*adj*) describes conditions which are greatly different from normal (↑) conditions, e.g. in northern countries, a room temperature of 45°C would be considered abnormal. **abnormality** (*n*).

exception (*n*) anything left out of a statement or description, e.g. the chlorides of the alkaline earth metals (p.117) are not hydrolyzed (p.66) in solution with the exception of beryllium and magnesium chlorides (these two chlorides are hydrolyzed). **except** (*v*).

exclude (*v*) to keep out, or to put out, e.g. (a) in the preparation of iron (II) sulphate crystals, atmospheric oxygen is excluded to prevent oxidation of the salt (the air is kept out of the apparatus); (b) in the reduction of copper (II) oxide by hydrogen, air is excluded by the hydrogen before heating takes place (hydrogen pushes the air out). **exclusion** (*n*), **exclusive** (*adj*).

excess (*n*) an amount that is more than the amount needed, e.g. if 100 cm³ of hydrochloric acid is the exact amount of acid needed to dissolve 5 g of zinc, then 150 cm³ of the acid is added to the zinc to make sure all the metal is dissolved. Excess acid has been added to the zinc, and there is an *excess* of 50 cm³ of hydrochloric acid. A volume of 150 cm³ of hydrochloric acid is 50 cm³ *in excess* of the 100 cm³ of acid needed to dissolve 5 g of zinc. **excessive** (*adj*).

intense (*adj*) describes a high degree (p.227) of a quantity, e.g. (a) an intense heat is at a very high temperature; (b) an intense radiation is a very powerful radiation. **intensity** (*n*).

converse (*n*) an equal and opposite action, or an equal and opposite statement, e.g. (a) condensation is the converse of evaporation, the two actions are equal and opposite; (b) the statement *'ideal gases obey Boyle's law'* has as its converse 'non-ideal gases do not obey Boyle's law' **converse** (*adj*).

introduce (*v*) to put a solid, a liquid, or a gas into a vessel (p.25) when skill is needed in the technique (p.43), e.g. to introduce a small quantity of a volatile liquid into a mercury barometer tube when measuring vapour pressure. **introduction** (*n*).

insert (*v*) to put an object in a fixed position, e.g. a thermometer in a cork.

excess   excess liquid in a flask

abundant (*adj*) describes something occurring
in great quantity, especially spread over a
large area, but which is not necessarily going
to be used, e.g. coal is abundant in Europe.

plentiful (*adj*) describes something which is in
great quantity and is available (p.85) for use,
e.g. (a) petrol is plentiful in Arabia (there is a
great quantity of it and it is put to use), (b) coal
burned in a plentiful supply of air forms carbon
dioxide and water. *Contrast limited* (↓).

100 cm³ alkali
neutralizes 100 cm³ acid

alkali
more than
100 cm³

sufficient
sufficient
alkali

alkali
100 cm³

neutralize

adequate
amount
of alkali

acid
100 cm³

adequate

adequate (*adj*) describes an amount that is
equal to the amount needed for a particular
purpose. The actual amount may not be
stated, but is assumed (p.222) to be known,
e.g. add an adequate amount of alkali to make
the mixture alkaline (the amount is not known).
adequacy (*n*).

sufficient (*adj*) describes an amount that is equal
to the amount needed for a particular purpose,
and is used with the name of the material or substance. To contrast *sufficient* and *adequate*:
there is *sufficient* acid to neutralize the alkali;
there is an *adequate* amount of acid to neutralize
the alkali. **sufficiency** (*n*), **suffice** (*v*).

insufficient (*adj*) describes a quantity that is not
sufficient (↑).

inadequate (*adj*) describes an amount that is
not adequate (↑) or anything that is needed
and is not adequate, e.g. (a) the amount of
acid was inadequate to neutralize the alkali;
(b) the voltage of the electrical supply was
inadequate to carry out the electrolysis; (c) his
knowledge of mathematics was inadequate for
his work in chemistry. **inadequacy** (*n*).

limited² (*adj*) (1) describes something which is
only available (p.85) for use in an inadequate
(↑) amount. It is the opposite of plentiful (↑),
e.g. coal burned in a limited supply of air forms
carbon monoxide and water. (2) describes the
application of theories and laws and the use of
instruments, devices and apparatus when
there are limits, e.g. (a) the law of constant
composition is limited to stoichiometrical
compounds (p.82); (b) ammeters have a
limited range of measurement.

deficient (*adj*) describes a material, object, or an idea that lacks a part or is lacking in quantity in a part of it, e.g. (a) if rubber is deficient in its sulphur content it will be too soft for many purposes; (b) the laboratory is deficient in fume cupboards. **deficiency** (*n*), **deficit** (*n*).

supplementary (*adj*) (1) describes an amount which is in addition to a previous (p.220) amount and is needed to complete or to improve a reaction or process. (2) describes an angle, which with another angle adds up to 180°, e.g. an angle of 124° is the supplementary angle of 56°. **supplement** (*n*), **supplement** (*v*).

apply (*v*) (1) to cause a force or a potential to act at a point or place, e.g. (a) to apply pressure to a gas in a vessel using a column of mercury; (b) to apply a voltage of 12 volts to the electrodes in a voltameter. (2) to use a theory or a law, e.g. Boyle's law is applied to all gases, but is only valid (↓) when applied to ideal gases (p.107). **application** (*n*).

valid (*adj*) describes a statement (p.222) or experiment that is accurate and is in agreement with scientific experience, e.g. the gas laws are valid for predictions of volume changes at low pressures.

relation (*n*) a connection: between quantities that can vary; between cause and effect (p.214); or between objects, e.g. (a) there is a relation between the mass of a metal deposited during electrolysis and the quantity of electric charge passed through the electrolyte; (b) there is a spatial (p.211) relation between the atoms in a molecule. **relate** (*v*), **related** (*adj*).

relative (*adj*) describes the relation (↑) between one physical property of a substance and the same physical property of a standard (p.229) substance, e.g. (a) relative vapour density is the ratio of the density of a gas divided by the density of hydrogen; (b) relative molecular mass is the ratio of the mass of one molecule of a substance divided by 1/12 of the mass of one atom of carbon-12.

supplementary

supplementary angle 124°

angle

56°

applying pressure to force water out of a ball

apply

respective (*adj*) describes the order relating objects to their descriptions when there are more than two such objects, e.g. the respective states of matter of oxygen, bromine and copper are gas, liquid and solid. **respectively** (*adv*).

correspond (*v*) to be similar (↓) in some part, function, structure or situation, without being identical, e.g. (a) the hydroxyl group in an alcohol corresponds to the hydroxyl group in water; (b) the nitrate (p.52) ion (p.123) and the carbonate ion (p.49) correspond in structure. **correspondence** (*n*).

similar (*adj*) two things are similar if they have many like properties, structures, qualities or characteristics, but have properties, structures, qualities or characteristics that distinguish (p.224) them. For example the characteristics of the transitional metals (p.121) are similar, yet each metal can be distinguished from the others. **similarity** (*n*).

identical (*adj*) describes two things which have the same number of properties and characteristics and these properties and characteristics are exactly the same. It is not possible to distinguish between identical objects except by the space they occupy (↓).

direct (*adj*) (1) describes a relation in which an increase in one quantity is related to the increase in a related quantity. (2) describes an inference (p.43) in which the properties of a subset are known and the inference is that the properties of the whole set will be the same.

inverse (*adj*) (1) describes a relation in which an increase in one quantity is related to a decrease in another quantity. (2) describes an inference (p.43) in which the properties of a whole set are known and the inference is that the properties of a subset will be the same.

occupy (*v*) to be in a particular place so that another object cannot be in that place, e.g. (a) the ions in a crystal (p.91) occupy positions in the lattice (p.92); (b) the alkali metals (p.117) occupy places in group I of the periodic system (p.119). **occupation** (*n*), **occupant** (*n*), **occupied** (*adj*).

correspond

the black triangles — are in corresponding positions in A, B and C

similar

A, B are similar

identical

A, C are identical

occupy

the green colour occupies the same position in A, B, C

International System of Units (SI)

PREFIXES FOR SI UNITS

MULTIPLE	FIGURE	PREFIX	SYMBOL
10^{12}	1 000 000 000 000	tera	T
10^9	1 000 000 000	giga	G
10^6	1 000 000	mega	M
10^3	1 000	kilo	k
10^{-3}	0.001	milli	m
10^{-6}	0.000 001	micro	μ
10^{-9}	0.000 000 001	nano	n
10^{-12}	0.000 000 000 001	pico	p

BASIC UNITS

Système International d'Unités (SI) is based on seven basic units; the seventh, the candela, is not included in the list below, as it is not used in chemistry.

metre *(unit of length)*   *symbol:* m
Defined from a wavelength in the spectrum of krypton.

kilogram *(unit of mass)*   *symbol:* kg
The mass of the international prototype kept at Sèvres.

second *(unit of time)*   *symbol:* s
Defined from a frequency in the spectrum of the caesium-133 atom.

kelvin *(unit of temperature)*   *symbol:* K
The fraction 1/273.16 of the temperature of the triple point of water.

mole *(unit of amount)*   *symbol:* mol
The amount of substance which contains as many elementary units as there are atoms in 0.012 kilogram of carbon-12. The elementary unit must be specified and may be an atom, a molecule, an ion, an electron, a radical, etc.

ampere *(unit of electric current)*   *symbol:* A
Defined from a newton and a metre.

DERIVED UNITS

QUANTITY	SYMBOL FOR QUANTITY	UNIT	SYMBOL FOR UNIT	BRIEF DEFINITION
velocity	v	m/s	—	displacement/time
acceleration	a	m/s^2	—	velocity/time
force	F	newton	N	mass × acceleration
energy	E	joule	J	force × distance
pressure	p	pascal	Pa	force/unit area
density	ρ	kg/m^3	—	kilogram/cubic metre
frequency	f	hertz	Hz	number of oscillations/time
concentration	—	mol/dm^3	M	moles/cubic decimetre
electric charge	Q	coulomb	C	amperes × time
electric potential	V	volt	V	joule/coulomb
electric resistance	R	ohm	Ω	volt/amperes
volume	V	dm^3	—	cubic decimetre

Symbols used in chemistry

LETTER/SYMBOL	QUANTITY, OBJECT OR OPERATOR
A	mass number. A_r relative atomic mass
E	energy, electromotive force
e	electron. $^0_{-1}$e charge and mass of electron
F	Faraday constant
f	frequency
H	heat of reaction
I	electric current
k	a constant
L	Avogadro constant
M	concentration in moles per cubic decimetre M_r relative molecular mass
m	mass
N	number of molecules
N	neutron number
n	any number; mole fraction; number of moles
n	a neutron. $^1_0$n charge and mass of neutron
p	pressure
p	a proton. $^1_1$p charge and mass of proton
Q	quantity of electric charge
R	molar gas constant, resistance
r	gas constant, radius
T	thermodynamic temperature (measured in kelvin)
t	time. $t_{\frac{1}{2}}$ half life
V	volume, electric potential difference V_m molar volume
Z	atomic number
$\triangle$	a change, e.g. $\triangle H$ change in heat
θ	temperature difference, temperature (Celsius scale)
ρ	density

Important constants

s.t.p. = standard temperature and pressure which is 1.00 atmospheres or 760 mm of mercury or 101 kPa and 273 K or 0°C.

Temperature of the triple point of water is 273.16 K.

Absolute zero of temperature is 0 K or − 273°C.

Volume of 1 mole of gas at s.t.p. is 22.4 dm^3 (molar volume).

Molar gas constant, 8.314 $J K^{-1} mol^{-1}$.

Avogadro constant, L = $6.02 \times 10^{23} mol^{-1}$.

Faraday constant, F, $9.65 \times 10^4 C mol^{-1}$.

Mass of electron, 9.11×10^{-31} kg.

Ratio of masses, proton/electron, 1840.

Ratio of masses, neutron/electron, 1840.

Charge on an electron, 1.6×10^{-19} C.

Ionic product of water $K_w = 1.008 \times 10^{-14} mol^2 dm^{-6}$ (298 K or 25°C).

1 calorie = 4.18 joules.

1 electron-volt (1 eV) = 1.6×10^{-19} J.

Common alloys

NAMES OF ALLOY	APPROXIMATE COMPOSITION	USES
brass	zinc 35% – 10%, copper 65% – 90%	decorative metal work
bronze		
– common	zinc 2%, tin 6%, copper 92%	machinery, decorative work
– aluminium	aluminium 10%, copper 90%	machinery castings
– coinage	zinc 1%, tin 4%, copper 95%	coins
dentist's amalgam	copper 30%, mercury 70%	dental fillings
duralumin	magnesium 0.5%, manganese 0.5%, copper 5%, aluminium 94%	framework of aeroplanes
gold		
– coinage	copper 10%, gold 90%	coins
– dental	silver 28%–14%, copper 14%–28%, gold 58%	dental fillings
lead, battery plate	antimony 6%, lead 94%	accumulators
manganin	nickel 1.5%, manganese 16%, copper 82.5%	resistance wire
nichrome	chromium 20%, nickel 80%	heating elements, resistance wire
pewter	lead 20%, tin 80%	utensils
silver		
– coinage	copper 10%, silver 90%	coins
solder	tin 50%, lead 50%	joining iron objects
steel		
– stainless	nickel 8%–20%, chromium 10%–20%, iron 80%–60%	utensils
– armour	nickel 1%–4%, chromium 0.5%–2%, iron 98%–95%	armour plating
– tool	chromium 2%–4%, molybdenum 6%–7%, iron 95%–90%	tools

Common abbreviations in chemistry

abs.	absolute	hyd.	hydrated
anhyd.	anhydrous	i.e.	that is
approx.	approximately	insol.	insoluble
aq.	aqueous	liq.	liquid
b.p.	boiling point	max.	maximum
conc.	concentrated	min.	minimum
concn.	concentration	m.p.	melting point
const.	constant	p.d.	potential difference
crit.	critical	ppt.	precipitate
cryst.	crystalline	r.a.m.	relative atomic mass
d.	decomposed	r.m.m.	relative molecular mass
decomp.	decomposition	sol.	soluble
dil.	dilute	soln.	solution
dist.	distilled	sp.	specific
e.g.	for example	s.t.p.	standard temperature and pressure
e.m.f.	electromotive force		
eqn.	equation	temp.	temperature
expt.	experiment	vac.	vacuum
fig.	figure (diagram)	v.d.	relative vapour density
f.p.	freezing point	wt.	weight
		°	degree (Celsius)

Understanding scientific words

New words can be made by adding **prefixes** or **suffixes** to a shorter word. Prefixes are put at the front of the shorter word and suffixes are put at the back of the word. Words can also be broken into parts, each of which can have a meaning, but cannot be used alone.

(i) correct → *in*correct   (adding a prefix)
   correct → correct*ness*   (adding a suffix)
   correct → *in*correct*ness*   (adding a prefix and a suffix)

(ii) **isomorphism** is broken into iso-morph-ism
   iso- is a prefix which means 'identical in structure'
   morph is a word part which means 'form or shape'
   -ism is a suffix which means 'a condition'

Hence *isomorphism* means the condition of having identical forms or shapes; it describes the condition of two crystalline substances.

Prefixes describing numbers or quantities are taken from Greek or Latin words. The following table shows the common prefixes from these two languages. Prefixes, suffixes, and word parts are listed alphabetically in separate sections after the table.

	GREEK PREFIX	LATIN PREFIX	PREFIX	MEANING	
1	mono-	uni-	hemi-	half	Gr
2	di-	bi-	semi-	half	L
3	tri-	ter-	poly-	many	Gr
4	tetra-	quad-	multi-	many	L
5	penta-	quinq-	omni-	all	L
6	hexa-	sex-	dupli-	twice	L
7	hepta-	sept-	tripli-	three times	Gr
8	octo-	oct-	hypo-	less, under	Gr
9	nona-	novem-	hyper-	more, over	Gr
10	deca-	deci-	sub-	under	L
100	hecta-	centi-	super-	over	L
1000	kilo-	milli-	iso-	same, equal, identical	Gr

PREFIXES

a-   without, lacking, lacking in, e.g. *a*morphous, being without shape; *a*symmetrical, without symmetry, or lacking in symmetry.

allo-   different, or different kinds, e.g. *allo*tropy, the existence of an element in two or more different forms.

amphi-   on both sides, e.g. *amph*oteric, having the nature of both an acid and a base.

an-   the same prefix as **a-**, used in front of words beginning with a vowel, or the letter *h*, e.g. *an*isotropic, not having the same properties in all directions; *an*hydrous, being without, or lacking, water, in a crystal.

anti-   opposite in direction, or in position, e.g. *anti*catalyst, a catalyst which slows down a chemical reaction, i.e. works in the opposite direction to a catalyst.

auto-   caused by itself, e.g. *auto*xidation, reaction of a substance with atmospheric oxygen at room temperature, the substance oxidizes itself; *auto*catalysis, a chemical reaction in which the products act as catalysts for the reaction.

cis-   on the same side, e.g. *cis*-compound, an isomer in which two like groups are on the same side of the double bond in the compound. See *trans-*.

co-   acting together, with, e.g. *co*hesion, the force holding two or more objects together.

counter-   acting against, acting in the opposite direction, e.g. *counter*act, to act against, such as a mild alkali counteracts the effect of acid on skin; *counter*clockwise, turning in the opposite direction to the hands of a clock.

de-   opposite action, e.g. *de*compression, the lessening of a pressure, it is the opposite action to compression; *de*activate, to make less active, it is the opposite of activate.

dia-   through, across, e.g. *dia*meter, the line going across a circle, through the centre.

dis-   opposite action, e.g. *dis*charge, to take an electric charge away from a charged body, the opposite of charge; *dis*connect, to break, or open, a connection, the opposite of connect.

equi-   having the same number, equal, e.g. *equi*molecular, having the same number of molecules; *equi*librium, the condition of two rates of reaction being equal and opposite, so that there is no further change in a reversible reaction.

im-   the opposite, not. (Used with words beginning with b, m, p.) For example, *im*perfect, not perfect, the opposite of perfect; *im*permeable, not permeable.

in-   the opposite, not. (Used with all words other than those beginning with b, m, p.) For example, *in*active, the opposite of active; *in*adequate, not adequate.

infra-   below, e.g. *infra*molecular, having a size smaller than a molecule, so the size is below molecular size.

inter-   between, among, e.g. *inter*face, a common surface between two liquids or two solids; *inter*stice, a narrow space between two solid objects.

macro-   great, large, e.g. *macro*molecule, a large molecule composed of many smaller molecules, as in a polymer.

micro-   small, especially if too small to be seen by the human eye alone, e.g. *micro*balance, a balance used for measuring masses of less than 1 mg; *micro*analysis, analysis using very small amounts of substances.

non-   not, e.g. *non*-electrolyte, a substance which is not an electrolyte; *non*-ferrous, any metal other than iron.

ortho-   straight, right-angled, upright, e.g. *ortho*gonal, with parts at right-angles; *ortho*rhombic, a crystal system with three unequal axes at right-angles.

pan-   all, complete, every, e.g. *pan*chromatic, covering all wavelengths of light in the spectrum.

para-   at the side of, by, e.g. *para*casein, an insoluble form of casein, formed when soluble casein coagulates.

pseudo-   has the same appearance, but is false, e.g. *pseudo*alum, a substance which has the appearance of an alum, but is not an alum.

re-   again, e.g. *re*activate, to make something activated again; *re*crystallize, to crystallize again.

syn, sym-   joined together, united, e.g. *syn*thesis, combining elements or compounds to make new compounds.

trans-   across, on the opposite side of, e.g. *trans*-compound, an isomer in which two like groups are on opposite sides of the double bond in the compound. See *cis-*.

ultra-   beyond, e.g. *ultra*filter, a filter which has holes so small it filters out colloids; it thus has uses beyond those of the ordinary filter.

un-   not, the opposite, e.g. *un*saturated, means not saturated; *un*stable, means not stable; *un*paired, means not in a pair, and so by itself.

SUFFIXES

-able   forms an adjective which shows an action can possibly take place, e.g. change*able*, something which can change; transform*able*, something which it is possible to transform.

-al   of, or to do with; forms a general adjective, e.g. experiment*al*, of, or to do with, experiment; fraction*al*, of, or to do with, fractions; therm*al*, of, or to do with, heat.

-ed   forms the past participle of a verb, can be used as an adjective; it shows an action under the control of an experimenter, e.g. vari*ed*, describes a quantity changed by an experimenter; dehydrat*ed*, describes a substance from which water has been removed under the control of an observer.

-er (-or)   forms a noun from a verb and describes an agent, e.g. mix*er*, a device which mixes; desiccat*or*, a device that desiccates; generat*or*, a device that generates a gas.

-gram   forms a noun describing a record which is written or drawn, e.g. chromato*gram*, the recorded result from an experiment on chromatography; tele*gram*, the written message recorded by telegraph.

-graph   forms a noun describing an instrument or device that records variation in a quantity, or other information, e.g. thermo*graph*, a kind of thermometer which records changes of temperature over a period of time; tele*graph*, a device which records information in words.

-ic   of, or to do with; forms a general adjective, e.g. bas*ic*, of, or to do with, a base; cycl*ic*, of, or to do with, a cycle; ion*ic*, of, or to do with, ions.

-ify   forms a verb which is causative in action, e.g. pur*ify*, to cause to become pure; solid*ify*, to cause to become solid.

-ing   forms the present participle of a verb, can be used as an adjective; it shows an action not under the control of an experimenter, e.g. fluctuat*ing*, describes a quantity varying above and below an average value, which cannot be controlled by an observer; disintegrat*ing*, describes a radioactive substance undergoing disintegration, as the process cannot be controlled by an observer.

-ity   forms a noun of a state or quality, e.g. pur*ity*, the quality or state of being pure; acid*ity*, the quality or state of being acid.

-ive   forms an adjective by replacing *-ion* in nouns; the adjective describes an agent producing the effect described by the noun, e.g. inhibit*ion* → inhibit*ive*, describes an agent causing inhibition; oxidat*ion* → oxidat*ive*, describes a process causing oxidation; explos*ion* → explos*ive*, describes an agent causing an explosion.

-ize   forms a verb which is causative in the formation of something, e.g. ion*ize*, to cause ions to be formed; polymer*ize*. to cause polymers to be formed.

-lysis   forms a noun describing the action of breaking down into simpler parts, e.g. hydro*lysis*, the decomposition of a compound by the action of water; electro*lysis*, the decomposition of a substance by an electric current.

-meter   forms a noun describing an instrument which measures quantitatively, e.g. thermo*meter*, an instrument which measures temperature accurately; volt*meter*, an instrument which measured electric potential in volts.

-metry   forms a noun describing a particular science of accurate measurement, e.g. thermo*metry*, the science of measuring temperature; hydro*metry*, the science of measuring the density of liquids.

-ness   forms an abstract noun of state or quality, e.g. sweet*ness*, the quality of being sweet; soft*ness*, the quality of being soft.

-ous   forms an adjective showing possession, or describing a state, e.g. anhydr*ous*, being in the state of not possessing water, homolog*ous*, in the state of being a homologue; homogen*ous*, in the state of having the same properties throughout a substance.

-philic   forms an adjective describing a liking for something, e.g. proto*philic*, describes a substance which accepts protons.

-phobic   forms an adjective describing a dislike for something, e.g. lyo*phobic*, describes a colloid which does not go readily into solution.

-scope   forms a noun describing an instrument which measures qualitatively, e.g. spectro*scope*, an instrument by which spectra can be observed qualitatively; hygro*scope*, an instrument which measures qualitatively the humidity of the atmosphere.

-scopy   forms a noun describing the use of instruments for observation in science, e.g. micro*scopy*, the use of microscopes for scientific observation.

-stat   forms a noun describing a device which keeps a quantity constant, e.g. hydro*stat*, a device which keeps water in a boiler at a constant level; thermo*stat*, a device which keeps a liquid, or an object, at a constant temperature.

-tion   forms an abstract noun. With -*ation*, it forms a noun of action, e.g. pollu*tion*, the result of polluting; concentra*tion*, the degree to which a solution is concentrated; distill*ation*, the noun of action from distil; precipit*ation*, the noun of action from precipitate.

WORD PARTS

aqua   water, to do with water, e.g. *aqu*eous, a solution containing water; *aqua*ion, an ion with molecules of water associated with it.

chrom   colour, to do with colour, e.g. pan*chrom*atic, all the colours, and hence all the wavelengths of the visible spectrum; *chrom*atography, the analysis of complex substances in which a coloured record of the analysis is produced.

gen   to produce, e.g. homo*gen*ize, to make a mixture of solid and liquid substances into a viscous liquid of the same texture throughout; *gen*erate, to produce energy or a flow of gas.

hydr   water or liquids, e.g. de*hydr*ate, to remove water; an*hydr*ous, describes a substance without water.

hygro   damp or humid, e.g. *hygro*scopic, attracting water from the atmosphere to become damp; *hygro*meter, an instrument that measures the relative humidity of the atmosphere.

morph   shape or form, e.g. a*morph*ous, describes a substance which is without a crystalline form; poly*morph*ism, existing in different forms.

photo   light, e.g. *photo*lysis, decomposition caused by light; *photo*halide, any halide which is decomposed by light.

pneumo   air or gas, e.g. *pneum*atic trough, a trough for the collection of gases.

pyro   great heat, e.g. *pyro*lysis, decomposition caused by heating; *pyro*meter, a special kind of thermometer for measuring very high temperatures.

therm   heat, e.g. *therm*ostable, stable when heated; *therm*al, of, or to do with heat; *therm*ometer, an instrument for the quantitative measure of temperature.

Index